Technology, Peace and Security | Technologie, Frieden und Sicherheit

Series Editor

Christian Reuter, Darmstadt, Germany

Technology plays a crucial role in various aspects of peace and security. This book series explores the intersection of computer science with peace and security studies. With a focus on cyber security and privacy, human-computer interaction as well as peace and conflict studies, it addresses topics such as peace informatics and technical peace research (cyber war, peace, arms control, and dual use), crisis informatics and information warfare (social media and collaborative technologies in conflict and crisis situations, misinformation and opinion manipulation), as well as usable safety, security, and privacy (resilient digital infrastructures, security and privacy enhancing technologies).

Technologie spielt eine entscheidende Rolle in verschiedenen Aspekten von Frieden und Sicherheit. Diese Buchreihe befasst sich mit Fragen an der Schnittstelle der Informatik mit der Friedens- und Sicherheitsforschung. Mit einem Fokus auf Cybersicherheit und Privatheit, Mensch-Computer-Interaktion sowie Friedens- und Konfliktforschung werden Themen wie Friedensinformatik und technische Friedensforschung (Cyberkrieg, Frieden, Rüstungskontrolle und Dual-Use), Kriseninformatik und Informationskrieg (soziale Medien und kollaborative Technologien in Konflikt- und Krisensituationen, Desinformation und Meinungsmanipulation) sowie nutzbare Sicherheit und Privatheit (resiliente digitale Infrastrukturen, Technologien zur Verbesserung der Sicherheit und des Datenschutzes) behandelt.

Philipp Kühn

Proactive Cyber Threat Intelligence

Automating the Intelligence Cycle Based on Open Sources

Springer Vieweg

Philipp Kühn
Science and Technology for Peace and Security
Technical University of Darmstadt
Darmstadt, Germany

ISSN 3004-9318 ISSN 3004-9326 (electronic)
Technology, Peace and Security | Technologie, Frieden und Sicherheit
ISBN 978-3-658-49240-3 ISBN 978-3-658-49241-0 (eBook)
https://doi.org/10.1007/978-3-658-49241-0

This Springer Vieweg imprint is published by the registered company Springer Fachmedien Wiesbaden GmbH, part of Springer Nature.
The registered company address is: Abraham-Lincoln-Str. 46, 65189 Wiesbaden, Germany

In an information-rich world, the wealth of information means a dearth of something else: a scarcity of whatever it is that information consumes. What information consumes is rather obvious: it consumes the attention of its recipients.

Hence a wealth of information creates a poverty of attention and a need to allocate that attention efficiently among the overabundance of information sources that might consume it.

– Herbert A. Simon, 1971

Foreword

With the ever-evolving threat landscape, securing IT infrastructure is a continuous challenge. Approaches like Cyber Threat Intelligence (CTI) sharing and Security Information and Event Management systems help detect attack campaigns but often require prior breaches to provide actionable insights. Meanwhile, valuable discussions about emerging threats exist in various sources, such as open-source software repositories, security advisories, blogs, and even the dark web. However, manually monitoring and analyzing these sources is time-consuming, prone to information overload, and lacks structured analytical capabilities.

The doctoral thesis of Philipp Kühn addresses these challenges by exploring automated approaches to extract cybersecurity-relevant information from both structured and unstructured sources to enhance Cyber Situational Awareness. By leveraging methods from Natural Language Processing, machine learning (ML), and data mining, it presents several key contributions to the CTI field. These include a specialized large language model (LLM) designed to improve NLP-based tasks such as classification, a taxonomy for CTI information, a prototype for identifying novel threat intelligence sources, and a qualitative analysis of the dark web.

The doctoral thesis of Philipp Kühn addresses important questions and provides meaningful findings. The unique contribution is the exploration and improvement of automated approaches to extract cybersecurity-relevant information from both structured and unstructured sources to support CERTs. Based on his contributions, Philipp Kühn was significantly involved as the first author and co-author of a total of 16 publications during his doctorate. The cumulative doctoral thesis contains seven papers.

Besides working on his thesis, Philipp has been actively involved in the supervision of student theses and teaching, *e.g.*, in the research group's programming courses and directing multiple programming groups. By spearheading the introduction of LaTeX templates and classes, he significantly streamlined the process for Bachelor's, Master's, and PhD students. He also took part in the group's IT team, consulting our IT administrator and developing our IT infrastructure further.

With his work, Philipp Kühn has proven that he is capable of independent scientific work. Thus, in March 2025 his dissertation was accepted by the Department of Computer Science at the Technical University of Darmstadt for the degree of *Dr.-Ing.* as the 11th PhD thesis in our research group PEASEC. I would like to see a further focus on topics of such high importance. Philipp, thank you for your contribution and for allowing me to accompany you to your PhD. I wish you all the best and every success for the future.

Prof. Dr. Dr. Christian Reuter
Science and Technology for Peace
and Security (PEASEC)
TU Darmstadt, Germany

Acknowledgements

This work represents the culmination of support from numerous individuals who have shaped my academic journey. I extend my deepest gratitude to my advisor, *Prof. Dr. Dr. Christian Reuter*, whose invaluable guidance, unwavering support, and enriching discussions have profoundly influenced the course and completion of this dissertation. I am sincerely thankful to my co-referee, *Prof. Dr. Harald Baier*, for his insightful feedback and expert counsel throughout this research endeavor. My appreciation also extends to *Prof. Dr. Thomas Schneider* and *Prof. Dr. Iryna Gurevych* for their willingness to serve on the doctoral committee and contribute their expertise to this culminating academic milestone.

I extend my heartfelt gratitude to the entire team at the chair *Science and Technology for Peace and Security (PEASEC)*. The intellectual diversity and stimulating discussions within our group have been instrumental in my journey, and it has been a true privilege to work alongside such exceptional colleagues.

I am particularly indebted to *Dr. Franz Kuntke* for your generous time, support, and our invaluable dialogues (and I am sorry for my impromptu Neovim tutorials), *Dr. Markus Bayer* for your patience, professional guidance and support, and enlightening discussions, and *Dr. Steffen Haesler* for your steadfast support during challenging times. My sincere thanks to *Kilian Demuth* and *Frank Nelles* for patiently enduring my technological and writing enthusiasm, and to *Anja-Liisa Gonsior* and *Laura Guntrum* for bringing joy and levity to our workplace through our conversations and shared humor. I am deeply grateful to *Dr. Thomas Reinhold*, *Dr. Katrin Hartwig*, *Dr. Jasmin Haunschild*, *Dr. Stefka Schmid*, *Dr. Thea Riebe*, and *Dr. Marc-André Kaufhold* for their welcoming presence, encouragement, and enriching academic discourse. I would also like to acknowledge the valuable contributions of my student assistants, *Dilara Nadermahmoodi*, *Tobias Frey*, *Frank Walter*, and *Marc Wendelborn*, whose dedicated support has significantly enhanced this work. I thank *Dr. Falko Marx*, *Arlene Kühn*, and *Paula*

Meissner for their reviews of this work. Finally, I extend my appreciation to all anonymous reviewers for their constructive feedback.

In addition, I want to thank all my co-authors *Kyra Wittorf*, *Julian Bäumler*, *Moritz Kerk*, *David Relke*, *Mike Schmidt*, and *Ramin Shanehsaz* for their valuable contributions towards this dissertation. This dissertation has been made possible by the funding of the German Federal Ministry of Education and Research (BMBF) and the Hessian Ministry of Higher Education, Research, Science and the Arts within their joint support of the National Research Center for Applied Cybersecurity ATHENE, the BMBF in the project CYWARN (13N15407), and the high-performance computer Lichtenberg at the NHR Centers NHR4CES at TU Darmstadt and the state governments participating on the basis of the resolutions of the Joint Science Conference (GWK) for national high performance computing at universities.

Lastly, I wish to express my profound gratitude to my friends and family who have been my pillars of support throughout this journey. To my parents, *Sabine* and *Andreas*, and my sister *Anna*, your steady encouragement, love, and faith in my abilities have been a constant source of determination towards my goal. To *Larina*, *Cleo*, *Malou*, and *Juna*, thank you for being the patient listeners to my near endless stream of ideas. I am deeply grateful to all my friends who helped me discover my true passions and identity. Each and every person mentioned has played an integral role in bringing this dissertation to fruition.

Above all, my most profound gratitude goes to my wife, *Camilla*. Your bottomless support, patient endurance of my truant thoughts, and unshakeable presence and willingness to reflect my state of mind have been invaluable to me. Your deep love and most loyal presence, especially during the challenging times, gave me the courage and strength to persevere and reach for the seemingly unreachable. You have been my foundation, helping me rise to every challenge and accomplish what was necessary. I stand where I am today because of you. Thank you, from the depths of my heart.

Abstract

Securing IT infrastructure is a tremendous task. With ever-changing infrastructures and threat landscapes, there is a constant need to stay ahead of time. To cope with this problem different methods are implemented, *e.g.*, Cyber Threat Intelligence (CTI) sharing or Security Information and Event Management (SIEM) systems which indicate current attack campaigns and threats. However, they usually require prior breaches to yield actionable information and are otherwise just distractions which call for attention, while, the web already contains discussions of upcoming threats. This may be an issue in a mailing list or version control system of Open-Source Software (OSS), vendors' security advisories, blogs or forums, or a rally in the dark web. When embedding CTI into the intelligence service's intelligence cycle, there even remains a gap of necessary functionalities. This results in a constant search for current information, usually a tedious and time-consuming manual task, and underdeveloped analytical capabilities of CTI software. And to top it all of, when reliable sources are identified, the result is usually a constant stream of information, leading to information overload – while the search for new sources does not stop due to the web's dynamics.

This thesis explores automated methods to extract actionable cybersecurity related information from official (usually structured) and unofficial (usually unstructured) sources, to foster defensive decisions. While CTI sharing is mainly *reactive* (a prior breach of partnering organizations is necessary), this thesis strives to explore *proactive* elements, *i.e.*, how to systematically include *open sources* in the context of CTI to provide Cyber Situational Awareness (CSA). To do so, I make use of various methods from the research fields of Natural Language Processing (NLP), machine learning (ML), crisis informatics, and data mining.

My contributions for the field of CTI are three-fold and range from conceptual over information-based contributions to analysis-based ones. I proposed a large language model (LLM) for the field of CTI with the goal to ease NLP based downstream tasks like classifications, which surpassed the state-of-the-art. This is followed-up by a taxonomy for CTI information, a prototype implementation to identify novel CTI sources, and a qualitative dark web analysis. Both latter proposals show the trade-off of crawling openly available information sources. For CTI information the clear web usually suffices as the dark web implements technological hindrances like CAPTCHAs. My analysis of industry standard CTI sources outlines signs of bad information quality which can be improved by common ML methods. Those methods provide suggestions to experts, which guide them during vulnerability assessment, and works on openly available sources like blogs or security advisories. Information overload is addressed by clustering techniques to provide an initial overview for experts, achieving over 90 % dimensional reduction across datasets, with deep clustering reaching 0.88 homogeneity for security bug reports (SBRs) and hierarchical clustering attaining 0.51 for advisory data.

In conclusion, the automation of various tasks in the daily work of security experts is a viable proposition that could prevent overlooking critical information and the onset of occupational burnout. By automating monotone and repetitive tasks of their daily workload, it may also enhance the depth and complexity of their work.

Author's Publications

During the course of this dissertation 16 publications were published. The following 7 publications are included as chapters in Part II of this thesis:

1. **Kuehn, P.**, Bäumler, J., Kaufhold, M.-A., Wendelborn, M., Reuter, C., "The Notion of Relevance in Cybersecurity: A Categorization of Security Tools and Deduction of Relevance Notions," in *Workshop-Proceedings Mensch und Computer*, Darmstadt: Gesellschaft für Informatik, 2022. https://doi.org/10.18420/muc2022-mci-ws01-220
2. Bayer, M., **Kuehn, P.**, Shanehsaz, R., Reuter, C., "CySecBERT: A Domain-Adapted Language Model for the Cybersecurity Domain," *ACM Transactions on Privacy and Security*, Mar. 15, 2024. https://doi.org/10.1145/3652594

   | CORE A and IF 3.0
3. **Kuehn, P.**, Nadermahmoodi, D., Bayer, M., Reuter, C., "Bandit on the Hunt: Dynamic Crawling for Cyber Threat Intelligence," *arXiv*, Apr. 25, 2025. https://doi.org/10.48550/arXiv.2504.18375
4. **Kuehn, P.**, Wittorf, K., Reuter, C., "Navigating the Shadows: Manual and Semi-Automated Evaluation of the DarkWeb for Cyber Threat Intelligence," *IEEE Access*, vol. 12, pp. 118 903–118 922, 2024. https://doi.org/10.1109/ACCESS.2024.3448247

   | IF 3.4
5. **Kuehn, P.**, Bayer, M., Wendelborn, M., Reuter, C., "OVANA: An Approach to Analyze and Improve the Information Quality of Vulnerability Databases," in *ARES '21: Proceedings of the 16th International Conference on Availability, Reliability and Security*, ACM, 2021, p. 11. https://doi.org/10.1145/3465481.3465744

   | ICORE B

6. **Kuehn, P.**, Relke, D. N., Reuter, C., "Common vulnerability scoring system prediction based on open source intelligence information sources," *Computers & Security*, 2023. https://doi.org/10.1016/j.cose.2023.103286

   | IF 4.8
7. **Kuehn, P.**, Bayer, M., Frey, T., Kerk, M., Reuter, C., "Reducing Information Overload: Because Even Security Experts Need to Blink," in *Availability, Reliability and Security—ARES 2025 Workshops*, Ghent, Belgium: Springer Nature, 2025

   | ICORE B

The following 9 papers are not included in the thesis, although their findings are supplementary to it:

8. **Kuehn, P.**, Riebe, T., Apelt, L., Jansen, M., Reuter, C., "Sharing of Cyber Threat Intelligence between States," *S+F Sicherheit und Frieden / Peace and Security*, vol. 38, pp. 22–28, 2020. https://doi.org/10.5771/0175-274X-2020-1-22
9. Egert, R., Gerber, N., Haunschild, J., **Kuehn, P.**, Zimmermann, V., "Towards Resilient Critical Infrastructures—Motivating Users to Contribute to Smart Grid Resilience," *i-com—Journal of Interactive Media*, vol. 20, pp. 161–175, 2021. https://doi.org/10.1515/icom-2021-0021

   | CiteScore 3.4
10. Kaufhold, M.-A., Fromm, J., Riebe, T., Mirbabaie, M., **Kuehn, P.**, Basyurt, A. S., Bayer, M., Stöttinger, M., Eyilmez, K., Möller, R., Fuchß, C., Stieglitz, S., Reuter, C., "CYWARN: Strategy and Technology Development for Cross-Platform Cyber Situational Awareness and Actor-Specific Cyber Threat Communication," in *Workshop-Proceedings Mensch und Computer*, Bonn: Gesellschaft für Informatik, 2021. https://doi.org/10.18420/muc2021-mciws08-263
11. Riebe, T., Wirth, T., Bayer, M., **Kuehn, P.**, Kaufhold, M.-A., Knauthe, V., Guthe, S., Reuter, C., "CySecAlert: An Alert Generation System for Cyber Security Events Using Open Source Intelligence Data," in *Information and Communications Security (ICICS)*, 2021, pp. 429–446. https://doi.org/10.1007/978-3-030-86890-1_24

    | ICORE C
12. Basyurt, A. S., Fromm, J., **Kuehn, P.**, Kaufhold, M.-A., Mirabaie, M., "HelpWanted—Challenges in Data Collection, Analysis and Communication of Cyber Threats in Security Operation Centers," in *Proceedings of the International Conference on Wirtschaftsinformatik (WI)*, Nürnberg,

2022. [Online]. Available: https://aisel.aisnet.org/wi2022/it_for_development/it_for_development/20/

13. Riebe, T., **Kuehn, P.**, Imperatori, P., Reuter, C., "U.S. Security Policy: The Dual-Use Regulation of Cryptography and its Effects on Surveillance," *European Journal for Security Research*, 2022. https://doi.org/10.1007/s41125-022-00080-0
14. Reinhold, T., **Kuehn, P.**, Günther, D., Schneider, T., Reuter, C., "Ex-TRUST: Reducing Exploit Stockpiles With a Privacy-Preserving Depletion Systems for Inter-State Relationships," *IEEE Transactions on Technology and Society*, vol. 4, pp. 158–170, 2023. https://doi.org/10.1109/TTS.2023.3280356
15. **Kuehn, P.**, Schmidt, M., Bayer, M., Reuter, C., "ThreatCrawl: A BERTbased Focused Crawler for the Cybersecurity Domain"
16. Reuter, C., Franken, J., Reinhold, T., **Kuehn, P.**, Kaufhold, M.-A., Riebe, T., Hartwig, K., Biselli, T., Schmid, S., Guntrum, L., Haesler, S., "Informatik für den Frieden: Perspektive von PEASEC zu 40 Jahren FIfF," FIfF-Kommunikation, 2024. [Online]. Available: https://peasec.de/paper/2024/2024_Reuteretal_InformatikFuerFrieden_fiff.pdf

Contents

Part II Publications

Abbreviations

A	Availability Impact
AC	Attack complexity
API	Application programming interface
APT	Advanced Persistent Threat
AV	Attack Vector
BERT	Bidirectional Encoder Representations from Transformers
BoW	Bag of Words
BTC	Bitcoin
C	Confidentiality Impact
CAPEC	Common Attack Pattern Enumerations and Classifications
CAPTCHA	Completely Automated Public Turing test to tell Computers and Humans Apart
CERT	Computer Emergency Response Team
CI	Confidentiality impact
CL-LSTM	Cross-Lingual LSTM
CNN	Convolutional Neural Network
CNNVD	China National Vulnerability Database of Information Security
CPE	Common Platform Enumeration
CRF	Conditional Random Field
CSA	Cyber Situational Awareness
CSAF	Common Security Advisory Framework
CTI	Cyber Threat Intelligence
CVE	Common Vulnerabilties and Exposures
CVSS	Common Vulnerability Scoring System
CWE	Common Weakness Enumeration
CySec	Cybersecurity
DAPT	Domain-adaptive pre-training

DBSCAN	Density-Based Spatial Clustering of Applications with Noise
DDoS	Distributed Denial-of-Service
DICE-E	Darknet Identification, Collection, Evaluation with Ethics
DL	Deep Learning
DOM	Document Object Model
DSSM	Deep Structured Semantic Model
DT	Decision Tree
DTL-EL	Deep Transfer Learning for Exploit Labeling
DWF	Dark web forum
DWM	Dark web marketplace
ETH	Ethereum
FT	Fine-tuned
GPU	Graphics Processing Unit
GRU	Gated recurrent unit
GT	Ground Truth
GUI	Graphical User Interface
HaaS	Hacking as a Service
HS	Hidden Service
HTML	Hypertext Markup Language
HTTP	Hyper Text Transfer Protocol
I	Integrity Impact
IDS	Intrusion Detection System
IE	Information Extraction
II	Integrity impact
IoC	Indicator of Compromise
IoT	Internet of Things
IPS	Intrusion prevention systems
IQ	Information quality
IR	Information Retrieval
IRC	Internet-Relay-Chat
ISS	Impact Sub Score
LDA	Latent Dirichlet Allocation
LLM	large language model
LP	Logic Programming
LR	Logistic Regression
LSTM	Long short-term memory
MAB	Multi-armed bandit
MAE	Mean Absolute Error
MISP	Malware information and sharing platform

ML	Machine learning
MLP	Multi-Layer Perceptron
MSE	Mean Squared Error
NB	Naive Bayes
NER	Named-Entity Recognition
NLP	Natural Language Processing
NN	Neural network
NPM	Node Package Manager
NVD	National Vulnerability Database
OS	Oversampling
OSINT	Open-source intelligence
OSS	Open-Source Software
OVAL	Open Vulnerability and Assessment Language
OVANA	Overt Vulnerability source ANAlysis
PFR	Point Frequent Function
POS	Part-of-Speech
PQC	Post-Quantum Cryptography
PR	Privileges required
PT	Pre-trained
QA	Question Answering
QC	Quantum Computing
RAT	Remote Administration Tool
RE	Relation Extraction
RegEx	Reguläre Expressions
RF	Random Forest
RNN	Recurrent neural network
S	Scope
SBERT	Sentence-BERT
SBR	Security bug report
SEM	Security Event Management
SGD	Stochastic gradient descent
SGNS	Skip-Gram with Negative Sampling
SIEM	Security Information and Event Management
SIM	Security Information Management
SN	Software name
SNA	Social Network Analysis
SOC	Security Operations Center
SOCMINT	Social media intelligence
SV	Software version

SVM	Support Vector Machine
TF-IDF	Term Frequency-Inverse Document Frequency
TIP	Threat intelligence platform
Tor	The Onion Router
TTP	Tactic, technique, and procedure
UI	User interaction
URL	Uniform Resource Locator
USD	US-Dollar
VF	Vulnerable function
VP	Vulnerable path
WWW	World Wide Web
XMR	Monero

List of Figures

List of Tables

Part I
Synopsis

Introduction 1

1.1 Motivation and Problem Statement

Since the early 2000s, digitization has become an integral part of every industry, revolutionizing how organizations operate and interact with their environment. While this digital transformation has unlocked unprecedented efficiency and innovation, it has also brought significant risks [20]. One of the most prominent side effects has been the steady rise in cybersecurity incidents, with each passing year seeing an increase in the frequency and complexity of attacks [21]. This trend has turned cybersecurity into a top priority for businesses and governments alike, as they grapple with the growing threats to their digital assets and functioning [22, 23].

The complexity of modern infrastructures worsens the challenge of maintaining robust cybersecurity defenses. Organizations rely on distributed and sophisticated systems, making them vulnerable to a wide range of attacks. As a result, defending these infrastructures requires a layered, comprehensive approach that encompasses multiple strategies. Traditional methods, such as keeping systems up-to-date and managing vulnerabilities, remain important, but they are no longer sufficient on their own [24].

In response, modern cybersecurity strategies have evolved to include red-team and blue-team exercises, which simulate real-world attacks and defenses, along with the growing incorporation of Cyber Threat Intelligence (CTI) [25]. CTI enables organizations to react to incoming threat information of partnering organizations, providing a more strategic, intelligence-driven approach to cybersecurity [26]. While traditional cybersecurity efforts have focused heavily on maintaining system integrity and incident response, contemporary approaches emphasize the importance of preemptive threat detection and informed decision-making through CTI [24]. This

P. Kühn, *Proactive Cyber Threat Intelligence*, Technology, Peace and Security | Technologie, Frieden und Sicherheit,
https://doi.org/10.1007/978-3-658-49241-0_1

shift underscores the need for advanced and comprehensive defenses in today's increasingly hostile cyber environment.

Despite the evolution of cybersecurity practices, the integration of CTI into the workflows of Computer Emergency Response Teams (CERTs), Security Operations Centers (SOCs) presents significant operational challenges. As cyber incidents rise in frequency and sophistication, these teams are often overwhelmed by the sheer volume of tasks. Security experts, particularly in national CERTs, are frequently overworked or understaffed, creating situations where existing resources are stretched thin [27–29]. This issue is exacerbated by the manual nature of many tasks, which include incident detection, analysis, and response over a multitude of different sources. For security services the framework of an intelligence cycle provides guidelines on which steps are necessary and build on each other, *i.e.*, the cycle is composed of the steps *Planning and Direction*, *Collection*, *Processing*, *Analysis and Production*, and *Dissemination*. Current research on this topic with regard to CTI show, it is yet not fully developed for CTI and lacks crucial functionalities in the steps *Collection*, *Processing*, and *Analysis* [30, 31].

While tools for managing CTI exist, they do not not fully address the specific needs of security experts in this domain, as [27, 28, 30, 31] noted. Feedback from these teams has revealed several critical shortcomings in existing tools [28]. Specifically, the lack of automated collection of threat information from unstructured sources, the ability to tailor this information to an organization's unique infrastructure, prioritization of incidents based on urgency, and deeper analytical capabilities are missing [32].

This presents significant barriers to efficient threat management. The absence of automated news collection forces staff to manually search for relevant intelligence, consuming valuable time and resources. Without proper tailoring, much of the information gathered is either irrelevant or lacks the context needed for effective decision-making [32]. Prioritization is another crucial element, as experts must decide which incidents demand immediate attention. Without reliable tools to assist in this, organizations risk overlooking critical threats. Lastly, deeper analytical capabilities are essential for understanding complex threats and responding appropriately, but these are often underdeveloped or unavailable in current automated solutions.

Therefore, while automation offers a potential solution to alleviate the workload of cybersecurity teams, the current tools fall short in addressing the full spectrum of needs [30]. The absence of these critical functionalities hinders the ability of CERTs and SOCs to effectively manage the growing number of incidents and take a proactive approach to cybersecurity, while their valuable time is used for mundane tasks. In the following, I will refer to information or data as raw data points, or an unprocessed collection of such, while intelligence is processed information which yields actionable information, *i.e.*, information that can be used for proactive cyber defense or new insights into attacker motives.

1.2 Aims and Objectives

Security experts must adapt to new technologies even before they are deployed in their infrastructure to be prepared for any cybersecurity incident. In addition, they must monitor every known information channel for recent cyber incidents on platforms in use to anticipate incoming threats. However, the volume of incoming information is overwhelming threefold [27, 28]:

Aspects of Information Overload in Cybersecurity [27, 28, 33]

1. Information is distributed in different sources, some of which may even be unknown.
2. The amount of information after pre-processing is still excessive.
3. Manually sifting through the information is time-consuming, repetitive, and reduces passion for the job and productivity.

While the former two aspects are methodical and technological problems, the latter is a social observation and must be monitored through via surveys or interviews. I will focus on the former two by focussing on the underlying data and algorithms, while aiming to answer the following research question:

Research Question
How to leverage open sources for an automated and proactive Cyber Threat Intelligence?

This research question can be divided into three sub-questions, which tackle *conceptual*, *information*-based, and *analytical* aspects:

RQ1 **What is cybersecurity information, how is it represented, measured by quality, and grasped by Cyber Threat Intelligence use-cases?**

Cybersecurity information is diverse and occurs in different sources. This increases the difficulty of gathering and normalizing every necessary bit of information for the field of CTI. Current research gathers a lot around social media [34] and anomaly detection [35]. Moreover, it requires an identification of current information sources, standards, and their use-cases. This research is complemented by an implementation of an cybersecurity large language model (LLM) to enable downstream tasks and an information metric for cybersecurity sources. The former is provided by a model, which lacks a comprehensive training set [36], while the latter is only partially defined in previous work [37–39].

| Adressed in Chaps. 7, 8, and 11.

RQ2 **What are efficient ways to automatically identify and find Cyber Threat Intelligence information in open sources?**

CTI information is scattered around the web and gathering the relevant bits is non-trivial [33, 40, 41]. There is the tip of the iceberg which might be known, but there is probably an abundance of unknown sources, which contain highly relevant information, *e.g.*, vulnerability write-ups and detailed vulnerability descriptions. Moreover, identifying the currently relevant information in either source is a tedious manual task, which requires time and consumes valuable workforce [27, 28]. A more efficient way is to let automated tools, like crawlers, sift through sources and alert operators in uncertain cases, while gathering and identifying the relevant information automatically [40]. They can scour the web to find new relevant cybersecurity sources (rather than just information), which makes the basis of information more robust and increase the certainty of operators. The efficiency of such an approach is measured by the harvest rate, *i.e.*, how much of the crawled information is actually relevant [40].

| Adressed in Chaps. 7 to 10.

RQ3 **How to analyze open information by unifying different inputs, removing overlapping information, and prioritizing the results to yield proactive Cyber Threat Intelligence?**

Automatic analysis of information becomes more important with rising numbers of data points and used sources. Otherwise, information overload would increase tremendously for the already overworked staff. This research question tackles the missing intelligence cycle phases of CTI, specifically, the phases *collection*, *processing*, *analysis*, and *dissemination* of intelligence [17, 30, 31]. The focus of research question are methods which improve the current state of the CTI cycle and support experts during their daily security work. This includes information quality analysis of sources, reduction of input data while keeping source information at hand, and vulnerability severity forecasting used for priorization, reports, and finally, the acutal security planning. I focus on data and applied methods rather than social and psychological aspects of their work. All these strategies are mandatory to (i) build automated pipelines and (ii) reduce the load of mundane tasks for security experts.

| Adressed in Chaps. 11 to 13.

1.3 Content and Structure of this Work

The remaining synopsis (*cf.* Part I) outlines further background information in Chap. 2, followed by the research design with my approach to tackle the research question, the corresponding details, and the context in Chap. 3. Chapter 4 presents my main findings with respect to the named questions (*cf.* Sect. 1.2), while Chap. 5 presents the answers to my questions, key contributions, limitations, and future work. Chapter 6 concludes this work.

Part II provides the empirical and theoretical findings to answer the research questions in different papers. Table 1.1 provides an overview of the different chapters aligned with the conceptual work, information gathering based work, and anal-

Table 1.1 Overview of the papers published in scope of this work. The categories are conceptualization (C), identification (I), and analysis (A). The ranking is based on numbers from 5. Februar 2025

	Chapter	Title	Venue	Ranking
C	Chap. 7	Relevance in Cybersecurity	MuC'22	–
	Chap. 8	CySecBERT	ACM TOPS	CORE A[a] IF 3.0
	Chap. 9	ThreatCrawl	arXiv	–
I	Chap. 10	Navigating the Shadows	IEEE Access	IF 3.4
A	Chap. 11	Overt Vulnerability Analysis	ARES'21	ICORE B
	Chap. 12	CVSS Prediction	C&S'22	IF 4.8
	Chap. 13	Reducing Information Overload	ARES CSA'25	ICORE B

[a]Based on the latest core journal ranking (2020).

ysis based work, which are all in line with the research questions (*cf.* Sect. 1.2). Chapter 7 presents the conceptual work based on a taxonomy and use-cases for security information, followed by a LLM implementation for the security domain, which is used in various of the following chapters. This is followed up by a crawler to identify clear web security information and sources, and a look at the dark web as CTI information source (*cf.* Chaps. 9 and 10). Chapters 11 to 13 propose means to analyze identified information for proactive CTI information, including different Natural Language Processing (NLP) operations, classification, and clustering (Table 1.1).

Chapter 7 *(The Notion of Relevance in Cybersecurity: A Categorization of Security Tools and Deduction of Relevance Notions)* presents a cybersecurity tool hierarchie with a correspesponding deduction of relevance notions for each category. It analyzes security tools and groups them into seven different categories. Related work for each groups is analyzed for the data used in (automatic) processing methods, to identify and present commonly used information types and sources. This chapter was originally published in the workshop proceedings of the *Mensch und Computer 2022—Workshopband: Usable Security und Privacy* [1].

Contributions and Highlights

- Overall taxonomy of security tools and sources
- Requirements of security experts for a cybersecurity dashboard

Author Statement It constitutes a joint work of Philipp Kühn, Julian Bäumler, Marc-André Kaufhold, Marc Wendelborn, and Christian Reuter. As corresponding and leading author, Philipp Kühn led the overall research design, management, and writing process of the paper. Julian Bäumler, Marc-André Kaufhold, and Philipp Kühn conducted a literature review (§2). Julian Bäumler and Marc-André Kaufhold build a categorization of security tools and concluded implications for a relevance term (§3). Philipp Kühn followed with the presentation of a relevance notion for each category (§4). Marc-André Kaufhold provided insights into a Computer Emergency Response Team dashboard (§5). The paper is concluded by Philipp Kühn (§6), including future work. Marc Wendelborn checked the related work, implemented the template restrictions, and proof-read the whole paper. Christian Reuter was a general advisor of this work and contributed with continuous feedback during all phases of the paper writing process.

CRediT Author-Statement [42] **Philipp Kühn:** Conceptualization, Methodology, Writing—Original Draft, Writing—Review and Editing. **Julian Bäumler:** Methodology, Visualization, Writing—Review and Editing. **Marc-André Kaufhold:** Methodology, Visualization, Writing—Review and Editing. **Marc Wendelborn:** Writing—Review and Editing. **Christian Reuter:** Resources, Supervision, Project administration, Funding acquisition.

Chapter 8 *(CySecBERT: A Domain-Adapted Language Model for the Cybersecurity Domain)* a cybersecurity domain-specific language model which is used as a building block in cybersecurity systems. It is evaluated on 15 extrinsic and intrinsic tasks, where it outperforms other models in domain-dependent scenarios. The model also shows that it is able to retain previously trained domain-independent knowledge. This chapter was originally published in the journal *ACM Transactions on Privacy and Security* [2].

Contributions and Highlights

- Outperforms state-of-the-art security LLM
- Retains general knowledge
- Enables different downstream tasks for the security domain

Author Statement It constitutes a joint work of Markus Bayer, Philipp Kühn, Ramin Shanehsaz, and Christian Reuter. Markus Bayer and Philipp Kühn led the overall research design, management, and writing process of the paper. The research methodology and experiments were initially designed and executed by Ramin Shanehsaz and validated by Markus Bayer and Philipp Kühn. Philipp Kühn carried out the data collection and filtering. Markus Bayer refined and extended the research methodology and experiments, and performed the analysis and interpretation of the results. Markus Bayer drafted the paper with contributions from Philipp Kühn. Both were involved in incorporating necessary revisions. Christian Reuter was a general advisor of this work and contributed with continuous feedback during all phases of the paper writing process.

CRediT Author-Statement **Markus Bayer:** Conceptualization, Methodology, Software, Validation, Formal analysis, Investigation, Data Curation, Writing—Original Draft, Writing—Review & Editing, Visualization, Project administration,

Funding acquisition. **Philipp Kühn:** Conceptualization, Methodology, Software, Validation, Data Curation, Writing—Original Draft, Writing—Review & Editing, Visualization, Project administration, Funding acquisition. Investigation **Ramin Shanehsaz:** Writing—Initial Draft, Software, Investigation, Visualization. **Christian Reuter:** Resources, Supervision, Funding acquisition.

Chapter 9 *(*Bandit on the Hunt: Dynamic Crawling for Cyber Threat Intelligence) addresses the challenge of automating the scanning process of multiple online portals and news pages to discover new threats for CTI using NLP. It proposes a new type of focused crawler that uses Bidirectional Encoder Representations from Transformers (BERT)-based models to classify documents while crawling to adapt its path dynamically. The study shows that the resulting crawler can successfully find relevant documents and modify its path accordingly, with harvest rates of up to 52%, which are better than the current state-of-the-art according to related work. However, the crawler has difficulties classifying the specific type of open-source intelligence (OSINT) named in texts, *e.g.*, Indicator of Compromise (IoC) content. This chapter was originally published arXiv [3].

Contributions and Highlights

- Comparison of different crawling methods.
- Identification of unknown security information sources.
- Outperforms state-of-the-art focused crawlers.
- Prototype of the system.

Author Statement It is a joint work of Philipp Kühn, Dilara Nadermahmoodi, Markus Bayer, and Christian Reuter. As corresponding and lead author, Philipp Kühn led the overall research design, implementation, data management, evaluation, and writing process of the paper. Dilara Nadermahmoodi supported the implementation and evaluation process. Markus Bayer reviewed the initial draft and help revise the paper. Christian Reuter was a general advisor for this work and contributed with continuous feedback at all stages of the paper writing process.

CRediT Author-Statement **Philipp Kühn:** Conceptualization, Methodoloy, Software, Validation, Data Curation, Writing – Original Draft, Writing – Review &

Editing, Visualization, Project administration. **Dilara Nadermahmoodi:** Software, Data Curation, Evaluation. **Markus Bayer:** Writing – Review & Editing. **Christian Reuter:** Resources, Supervision, Funding acquisition.

Chapter 10 (Navigating the Shadows: Evaluating the Dark Web for Cyber Threat Intelligence) aims to enhance Cyber Threat Intelligence (CTI) by analyzing the dark web as an Open Source Intelligence (OSINT) data source, addressing the gap left by the current focus on the clear web. Through a comprehensive literature review and manual analysis of 65 dark web forums, 72 marketplaces, and single-vendor-shops, the study identifies IT security-relevant information, such as malware toolboxes, and highlights challenges like anti-crawling measures. The findings suggest that while the dark web contains valuable CTI data, but requires specialized methods for effective extraction due to interaction barriers and should mainly be used, when capacity for this procedure remains. This chapter was originally published in the journal *IEEE Access* [4].

Contributions and Highlights

- Qualitative analysis of dark web marketplaces (DWMs) and dark web forums (DWFs)
- Identification of hindrances for automated crawling and potential mitigations
- Presentation of problems for CTI information gathering

Author Statement It constitutes a joint work of Philipp Kühn, Kyra Wittorf, and Christian Reuter. Philipp Kühn led the overall research design, management, and writing process of the paper. The research methodology and experiments were initially designed and executed by Kyra Wittorf and validated by Philipp Kühn. Kyra Wittorf carried out the data collection and filtering. Philipp Kühn refined and extended the research methodology, and performed the analysis and interpretation of the results, followed by the rewriting process of the paper as a whole. Philipp Kühn was involved in incorporating necessary revisions. Christian Reuter was a general advisor of this work and contributed with continuous feedback during all phases of the paper writing process.

CRediT Author-Statement **Philipp Kühn:** Conceptualization, Methodology, Validation, Data Curation, Writing—Original Draft, Review & Editing, Project administration. **Kyra Wittorf:** Conceptualization, Methodology, Data Curation, Writing—Initial Draft, Software, Investigation, Visualization. **Christian Reuter:** Resources, Supervision, Funding acquisition.

Chapter 11 *(OVANA: An Approach to Analyze and Improve the Information Quality of Vulnerability Databases)* examines the information quality of vulnerability databases, exemplified at the de facto standard vulnerability database National Vulnerability Database (NVD). This chapter defines an information quality metric and identifies several problems with the NVD based on this metric. Those problems are automatically tackled using machine learning (ML), which improves the NVD's information quality by more than 50%. This chapter was originally published in the conference proceedings of the *International Conference on Availability, Reliability, and Security* [5].

Contributions and Highlights

- Identification of core problems of the NVD's vulnerability descriptions and assessments
- Definition of information quality (IQ) metric for vulnerability databases
- Tools to support experts and improve IQ

Author Statement It constitutes a joint work of Philipp Kühn, Markus Bayer, Marc Wendelborn, and Christian Reuter. As corresponding and leading author, Philipp Kühn led the overall research design, management, and writing process of the paper. The authors conducted a literature review led by Philipp Kühn, with additions and a review process by Marc Wendelborn and Markus Bayer. Philipp Kühn and Markus Bayer designed the Research Methodology together. The fundamental implementation and data preparation was done by Marc Wendelborn, whereas the machine learning modules were implemented, trained, and evaluated by Markus Bayer. The results of the paper were written by Philipp Kühn, whereas the limitations and contributions were written equally by Markus Bayer and Philipp Kühn. Christian Reuter was a general advisor of this work and contributed with continuous feedback during all phases of the paper writing process.

CRediT Author-Statement Philipp Kühn: Conceptualization, Methodology, Software, Formal analysis, Investigation, Writing—Original Draft, Writing—Review and Editing, Visualization. **Markus Bayer:** Software, Validation, Formal analysis, Data Curation, Writing—Review and Editing **Marc Wendelborn:** Software, Writing—Review and Editing **Christian Reuter:** Resources, Supervision, Project administration, Funding acquisition.

Chapter 12 *(Common Vulnerability Scoring System Prediction based on Open Source Intelligence Information Sources)* focuses on the use of publicly available web pages referenced in the NVD as sources of text through web scraping to predict Common Vulnerability Scoring System (CVSS) vectors using Deep Learning (DL) based methods. The authors provide a classification of the NVD's reference texts based on the suitability and crawlability of their texts. The study shows that the use of web scraping and NVD techniques can improve the prediction of CVSS vectors compared to the state-of-the-art, but requires focusing on high quality information. Bad pre-selection of sources improves the classification process just negligibly. This chapter was originally published in the journal *Computers & Security* [6].

Contributions and Highlights

- Classification of vulnerability information sources in the NVD
- Tailored scrapers for information sources
- Comparison of different models for vulnerability severity predictions (CVSS) based on open sources

Author Statement It is a joint work of Philipp Kühn with David Relke and Christian Reuter. As corresponding and lead author, Philipp Kühn led the overall research design, management and writing process of the paper. Following the structure of the paper, Philipp Kühn wrote the introduction. Related Work was a collaborative work between David Relke and Philipp Kühn, where Philipp Kühn defined the research gap. David Relke carried out the preliminary analysis of the dataset and defined the crawling objectives under the guidance of Philipp Kühn. The grouping of information sources is based on this analysis. The collection of the dataset for the training process is done by David Relke. Philipp Kühn and David Relke trained the various CVSS prediction models. The evaluation was a joint work of David Relke and Philipp Kühn. The discussion was drafted by David Relke and refined by Philipp

Kühn. Philipp Kühn concluded the paper. Christian Reuter was a general advisor for this work and contributed with continuous feedback at all stages of the paper writing process.

CRediT Author-Statement Philipp Kühn: Conceptualization, Investigation, Methodology, Software, Writing—Review and Edition, Visualization. **David Relke:** Methodology, Software, Investigation, Writing—Initial Draft. **Christian Reuter:** Resources, Supervision, Project administration, Funding acquisition.

Chapter 13 *(Reducing Information Overload: Because Even Security ExpertsNeed to Blink)* addresses the challenge Computer Emergency Response Team (CERT)s facing critical operational challenges processing the growing volume of security-related information, which leads to burnout and attrition among security professionals due to manual analysis requirements. Our evaluation of 196 clustering algorithm and embedding model combinations across five security-related datasets demonstrates an approach for automated information consolidation with minimal configuration requirements. The results show that clustering can reduce information processing requirements by over 90% while maintaining semantic coherence, achieving homogeneity scores up to 0.88 for security bug reports and potentially saving over 3 750 work hours annually per analyst. This chapter was originally published in the workshop proceedings of the *International Conference on Availability, Reliability, and Security* [7].

Contributions and Highlights

- Creation of THREATREPORT, a novel labeled threat report corpus.
- Systematic evaluation of 196 clustering-embedding combinations demonstrating 90% information reduction while maintaining semantic coherence.
- Empirical validation showing deep clustering achieves 0.88 homogeneity for security bug reports (SBRs) and hierarchical-based clustering reaches 0.51 for advisories.

Author Statement It constitutes a joint work of Philipp Kühn, Markus Bayer, Tobias Frey, Moritz Kerk, and Christian Reuter, and was funded within the research project CYWARN and ATHENE SecUrban. As corresponding and leading author, Philipp Kühn led the overall research design, management, and writing process of the paper. He conceptualized the initial idea, whereas Moritz Kerk implemented an initial prototype of it. Philipp Kühn later refined the idea based on the findings with suggestions by Markus Bayer. Subsequenctly, Philipp Kühn implemented the new idea, gathered the data, wrote a full initial draft of the paper, and refined it based on Markus Bayer's feedback. Tobias Frey supported the implementation and bug-fixing process, as well as the evaluation. Christian Reuter was a general advisor of this work and contributed with continuous feedback during all phases of the paper writing process.

CRediT Author-Statement Philipp Kühn: Conceptualization, Methodology, Software, Validation, Invesitation, Data Curation, Writing—Original Draft, Review, & Editing. **Markus Bayer:** Writing—Review. **Tobias Frey:** Software. **Moritz Kerk:** Software, Data Curation. **Christian Reuter:** Resources, Supervision, Funding acquisition.

Research Background and Field 2

CTI combines a multitude of research fields, exceeding the span of computer science. I will however, focus on the core fields, that are touched by this work and expand on them in the following subsections. First, by its nature, CTI has been recognized as a subset of broader *intelligence tasks*, with its methodologies aligning closely with the intelligence cycle [17, 30, 31] (*cf.* Sect. 2.1). Second, through its methods and information which are usually unstructured natural texts, the fields *NLP* and *ML* are intertwined with CTI as well [2, 32] (*cf.* Sect. 2.2). The last field, which serves as background for this work is the field of *data mining* and *Information Retrieval (IR)*, which provides core methodologies to automate relevant aspects of the security cycle (*cf.* Sect. 2.3).

2.1 Cyber Threat Intelligence as Intelligence Task

Intelligence tasks originate in national intelligence services, which operationalized different aspects of their daily work. On of them is the so-called *intelligence cycle* which refers to a systematic process that involves the collection, analysis, dissemination, and utilization of information (*cf.* Fig. 2.1). Traditionally it is used in military and government contexts to support decision-making [17]. The original cycle consists of several distinct stages, each critical to turning raw data into actionable intelligence [17]. The cycle is depicted in Fig. 2.1 and comprises the stages:

Planning and Direction This involves identifying the specific intelligence requirements and objectives, guiding the entire process.

Collection The gathering of relevant information from various sources to meet the intelligence requirements.

P. Kühn, *Proactive Cyber Threat Intelligence*, Technology, Peace and Security | Technologie, Frieden und Sicherheit,
https://doi.org/10.1007/978-3-658-49241-0_2

Processing and Exploitation The collected information is processed, categorized, normalized, and prepared for analysis.

Analysis and Production Analysts evaluate the processed information to identify patterns, correlations, and insights, transforming the information into actionable intelligence.

Dissemination The intelligence is distributed to the relevant stakeholders, enabling them to take informed actions.

Feedback This final step involves gathering feedback from stakeholders to refine and improve future intelligence operations.

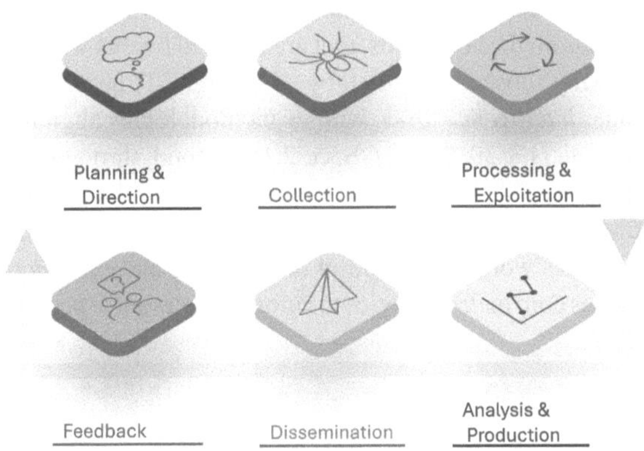

Fig. 2.1 The original intelligence cycle as defined by [17]. Together, all steps, starting at *Planning & Direction* and ending at *Feedback*, provide a framework for intelligence generation of intelligence services. The color-codes are used over the course of this synopsis

By viewing this cycle as part of CTI, organizations can adopt a structured approach to manage and interpret threat data, and enhance their overall cybersecurity posture [30]. Despite CTI's alignment with the intelligence cycle, research has identified key gaps that prevent CTI from fully leveraging the potential of this structured approach [30, 31]. According to Sauerwein et al. [30] key functionalities are either underdeveloped or absent from current CTI implementations, such that security professionals are hindered in their daily work [27, 28]. Many stages of the CTI cycle are still largely manual, particularly in the computable stages *Collection*, *Pre-processing*, and *Analysis*. Automation tools that could streamline these processes are either incomplete or missing altogether, which leads to information overload for experts or even burnout. While information collection is common, the ability to contextualize this information to a specific organization's infrastructure is often inadequate or not possible due to inconsistent or missing information [5, 39]. Existing systems struggle to filter out irrelevant information or prioritize on the relevant ones, making it difficult to focus on the most pressing threats [27]. Without clear guidance on which threats demand immediate attention, cybersecurity teams may spend valuable time on less critical issues like a daily and mostly manual gathering of information [28]. The speed of threat evolution often outpaces the ability of current CTI systems to analyze and disseminate intelligence. This lag creates a gap between the detection of a potential threat and the organization's ability to respond to it. For CTI, the intelligence cycle will retain the core principles of the original intelligence cycle, but addresses the specific needs for cybersecurity. The following descriptions provide an overview of the different stages with their functioning in CTI Fig. 2.2.

Planning & Direction Define specific cybersecurity objectives, such as identifying key assets to protect, the types of threats to monitor, and the most relevant sources of threat data (*e.g.*, open-source information, dark web monitoring, and incident reports), basically the scope of the intelligence one aims to produce.

Collection Collect relevant threat information from a variety of sources, including security feeds, malware reports, vulnerability databases, social media platforms where threat actors may communicate, and open sources like blogs or news outlets.

Pre-Processing Normalize, and prepare the collected threat data for analysis. This often includes enriching data through correlation with historical incidents or contextual information about adversaries.

Analysis Interpret the threat information to assess its relevance and potential impact on the organization's infrastructure. This phase transforms raw threat data into contextual intelligence, which can then be used to anticipate attacks or respond effectively to incidents.

Dissemination Intelligence is communicated to relevant cybersecurity teams, such as Security Operations Center (SOC) or Computer Emergency Response Team (CERT) personnel, decision-makers who can act upon it to mitigate risks, and (maybe) partner organizations or the public.

Evaluation & Feedback Continuous feedback is provided by the cybersecurity teams to ensure that the CTI process remains aligned with the organization's evolving security needs.

When it comes to computation-based methods used in the cycle, they stay in the realm of the stages *Collection*, *Pre-Processing*, and *Analysis*. The stage *Planning & Direction* is about identifying the scope of the upcoming intelligence generation. This can either be information about the infrastructure to gather threat information, the threat actor to monitor, or even the sources one is interested to use for the process. After all the collection, processing, and analysis, the *Dissemination* is ready, which is a more or less political or strategic decision, *i.e.*, which information should leave ones perimeter or even control. The *Evaluation & Feedback* is based on the resulting intelligence and how to improve the total CTI cycle. Sauerwein et al. [30] also include a phase they call *Cross-Process Support* which includes aspects like the security and privacy of the whole cycle, data quality measurements, build-up of trust, and collaborations. This phase is omitted in this version of the CTI cycle, as it is mainly out of scope of the present work. I focus on the computation-based stages in the cycle, as my main focus is the information, methods, and algorithms used to gather and process the information, with excursions into intelligence generation, *i.e.*, automating the mundane tasks which prevent professionals from working on what they applied for: *secure systems* and *generate Cyber Threat Intelligence*. The stages *Collection*, *Pre-Processing*, and *Analysis* are time-consuming and drain energy of experts [27, 28], while *Dissemination* is the work why experts usually get into the field [27], requires a broader understanding of all aspects of cybersecurity, and, political and strategic decision-making. Current algorithms lack this aspect and so they should at most support humans in this step. Figure 2.2 shows all stages of the CTI cycle outlined above, while the stages in focus of the present work are colored.

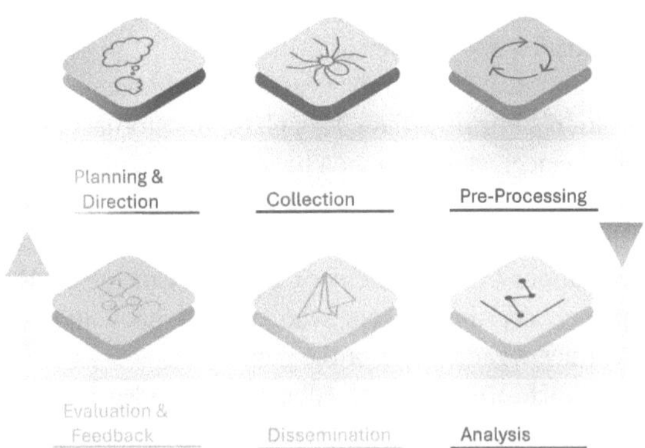

Fig. 2.2 The Cyber Threat Intelligence cycle. Grayed phases are not the focus of the present work, but may be touched, *e.g.*, *Dissemination* in Chap. 12

The field of CTI is usually applied in corporations in so called CTI platforms [43]. An example of such a platform is malware information and sharing platform (MISP), which houses several commodity services such as feed collection, aggregation, tagging incoming data, and more. This enables operators and experts to handle their tasks and reports in one single place, rather than a multitude of such. The present work evaluates and proposes multiple such services over the course of the whole CTI cycle, which complements such platforms with more ways for automation, *e.g.*, predicting vulnerability severity with automatic CVSS assessment methods. Besides, the CTI community hosts a lot of standards, *e.g.*, MITRE's ATT&CK framework. It provides a standardizes map of different adversarial behavior. Given a monitored behavior, this allows to map behavior to Advanced Persistent Threats (APTs) for example. It might hint in the involved nationalities and help with attack attribution. While this is an important component for CTI, it differs from the goals with in the present work, as it provides a way to map given information onto tactic, technique, and procedures (TTPs). The present work, however, supports this process by proposing novel automated ways to monitor sources, which serves as a early warning system. In addition, it can enrich existing ATT&CK techniques with more nuanced information and context, and improve the timeliness and relevance of ATT&CK updates with faster update cycles.

2.2 Natural Language Processing and Machine Learning

NLP and ML are fields focused on enabling machines to understand, interpret, and generate human language in a meaningful way [44] and analyze the resulting data to yield intelligence. In NLP, among other aspects, text data is converted into machine-readable representations, or embeddings, that encode semantic meaning, allowing to process language efficiently and perform various tasks like sentiment analysis, translation, and question answering [44, 45]. A significant breakthrough in NLP came with the development of transformers, which are deep learning models designed to handle sequences of data by capturing contextual relationships between words across long distances in a text [46]. Transformers serve as the backbone of LLMs that can generate coherent, contextually relevant responses to user prompts, summarize documents, and perform complex language tasks [45–47]. For example, embeddings created by Bidirectional Encoder Representations from Transformers (BERT) can represent sentences with fine-grained nuances [45, 48], enabling better search and IR [15, 49–51], while the GPT-based ChatGPT allows human-like conversations [52, 53].

The integration of ML in cybersecurity and CTI has become essential for dealing with the scale and complexity of modern threats, providing benefits across various domains such as threat detection, threat intelligence enrichment, incident response, and even policy automation [2, 54–56]. With the volume of data that security teams manage – from logs and alerts to network traffic and external intelligence – NLP and ML offer an effective solution to filter noise, identify potential threats, and generate actionable insights [57–60]. NLP techniques, for instance, are particularly beneficial in parsing and analyzing unstructured threat intelligence feeds, security reports, and social media posts, extracting relevant details like IoCs, TTPs, and exploited vulnerabilities [6]. This enables the automation of intelligence collection and reduces manual labor, allowing analysts to focus on higher-order tasks such as correlation and strategy development.

One of the most significant applications of ML in cybersecurity is in anomaly detection [58, 61, 62]. Anomaly detection identifies patterns that deviate from normal behavior, highlighting unusual activities that may indicate a problem, threat, or even current attacks [62]. By employing ML algorithms, such systems can continuously learn and adjust to network behavior, becoming more accurate over time and reduce false positives. For example, ML models can analyze user activity and access patterns, detecting anomalies that could signify credential misuse or privilege escalation attempts [58, 63, 64]. These insights allow organizations to respond swiftly to potential breaches and mitigate damage before incidents escalate.

Another critical benefit of ML and NLP in cybersecurity is the ability to predict emerging threats by analyzing large datasets from both internal logs and external threat intelligence sources [35, 57, 58]. Predictive models can highlight possible future attack vectors based on historical data and trends, enabling proactive defenses and allowing security teams to prioritize patches and mitigate vulnerabilities. NLP, specifically through advanced embeddings and transformers, allows for sentiment analysis and language pattern recognition across social media and forums where hackers may discuss vulnerabilities, allowing organizations to gain early warnings of potential attack trends [11, 65, 66]. For example, using embeddings to detect mentions of new exploits or vulnerabilities in hacker forums can provide actionable intelligence, resulting in preemptive defenses before the vulnerability is widely exploited [67, 68]. Simple keyword search would not suffice due to the dynamic naming strategies by threat actors [69].

The benefits of NLP and ML in cybersecurity extend to resource optimization and overall efficiency within security operations [27, 28]. Automating repetitive tasks like log analysis, phishing detection, and threat classification allows teams to address more sophisticated tasks and respond more rapidly to incidents. This not only improves the accuracy and speed of threat identification but also reduces the likelihood of alert fatigue, which is a major challenge in SOCs where analysts often face overwhelming volumes of alerts [27, 28]. By reducing false positives and focusing resources on higher-risk incidents, NLP and ML contribute to a more sustainable and strategic approach to cybersecurity. Furthermore, ML models can be used to dynamically adapt security policies based on the latest threat landscape, allowing for more adaptive defenses and a security posture that evolves in real-time [70–72].

2.3 Data Mining and Information Retrieval

Data mining is a field dedicated to discovering patterns, relationships, and valuable insights from large datasets, often through techniques such as clustering, classification, association, and anomaly detection [73, 74]. It leverages statistical methods and database systems to process complex datasets and uncover previously unknown correlations within the data (*cf.* Sect. 2.2). Data mining plays a key role in numerous fields, including business [75], healthcare [76], and cybersecurity [77], by transforming raw data into actionable intelligence that can inform decision-making. The core objective of data mining is to enable better predictions and decisions by uncovering trends, detecting irregularities, and generating a deeper understanding of data through automated analysis [78].

IR is the process of obtaining relevant information from large collections of unstructured or semi-structured data, with a focus on retrieving documents, texts, or records that meet a user's query [79]. It involves indexing, searching, ranking, and organizing data to allow efficient access to relevant content, often relying on techniques such as keyword matching, vector space models, and semantic analysis [79]. IR is widely used in search engines [80], recommendation systems [81, 82], and digital libraries [83], where the goal is to deliver the most relevant information based on a given input query. Key advances in IR include NLP and ML algorithms, which enhance retrieval accuracy and enable the identification of relevant information with high precision in extensive data collections (*cf.* Sect. 2.2). Current trends in IR leverage LLMs to improve the retrieval accuracy of knowledge bases through retrieval augmented generation [53].

In the context of cybersecurity, data mining and IR are essential for identifying patterns in network activity, threat intelligence feeds [84], and logs to detect threats [35, 58, 85], recognize vulnerabilities [86], and uncover suspicious activities [87, 88]. Data mining techniques like classification and clustering help in categorizing malware [89], identifying phishing attempts [59], and clustering malicious IP addresses to reveal coordinated attacks [90]. IR complements these efforts by parsing and extracting relevant data from massive sources, such as security databases [5], news feeds, and social media [11], to help analysts understand emerging threats and trends. For instance, retrieving relevant IoCs or parsing documents about recent vulnerabilities can significantly enhance an organization's response time and preparedness [41]. Together, data mining and IR contribute to faster detection, better threat intelligence, and improved situational awareness, all critical elements in strengthening cybersecurity defenses.

Current research in data mining and IR for cybersecurity focuses on developing advanced ML algorithms that can handle large-scale data, filter noise, and identify novel attack patterns in real time [91, 92]. Researchers are exploring methods like deep learning, graph-based analysis, and adversarial machine learning to improve detection accuracy and resilience against increasingly sophisticated cyber threats. Another area of focus is real-time threat intelligence aggregation, where data mining techniques are combined with IR to analyze OSINT and dark web data for early threat detection [41, 93–96]. Techniques like NLP-enhanced IR are also being researched to analyze text-heavy data sources, such as hacker forums and incident reports, automatically extracting insights that enhance threat anticipation and response capabilities.

3 Research Methodology and Context

This chapter presents the CTI intelligence cycle and research methodology, which are derived from the research questions posed at the outset of this study (*cf.* Sect. 1.2). The bigger picture of this research is presented in Sect. 3.2. I outline my contributions to the funding projects and how they complement other research in these projects.

3.1 Methodology

This section outlines how my methodology used in the following chapters and aligns with my research questions. I will reframe the intelligence cycle for the CTI domain, identify missing pieces to automate mundane steps, and propose my methodologies used in Part II to answer my research questions.

3.1.1 Challenges Buried in the CTI Cycle

The tasks involved in each stage of the CTI cycle (*cf.* Sect. 2.1) reflect the key challenges that CERTs and SOCs face in managing modern cybersecurity risks. These challenges fall into several categories, aligned with the CTI cycle stages *Collection*, *Pre-processing*, *Analysis*, and *Dissemination*:

Automated Collection of Threat Data Organizations typically rely on a fixed set of known information sources, with limited opportunity to expand or strengthen this foundation. The collection process often requires manual effort due to frequent changes made by vendors and information outlets in website designs. This

P. Kühn, *Proactive Cyber Threat Intelligence*, Technology, Peace and Security | Technologie, Frieden und Sicherheit,
https://doi.org/10.1007/978-3-658-49241-0_3

creates a significant challenge: how to automatically gather threat intelligence from diverse sources without human intervention—whether that's scraping news sites, monitoring security forums, or identifying new information sources.

| *Challenges:* Identification, Collection.

Data Preparation and Contextualization Security experts need to efficiently filter through vast amounts of data to identify the most relevant information for their organization's infrastructure and devices. This requires carefully aligning raw data with the organization's specific infrastructure and security requirements, which includes mapping potential vulnerabilities to particular assets and evaluating priority levels of different attack vectors.

The data must also be standardized into a consistent format before it can be effectively analyzed. This standardization is particularly crucial when working with unstructured natural language text, which is commonly used in security reports and advisories.

| Challenges: Structure, Filtering.

Incident Prioritization Even after appropriate pre-processing and analysis, security professionals still face information overload that impedes their work [28]. Therefore, it's essential to filter and prioritize security incidents based on their severity, relevance, and potential impact to ensure the most critical threats are addressed first.

| *Challenges:* Priorization, Enhancement.

In-depth Analysis and Reporting Comprehensive analysis of collected data is essential for identifying potential attack patterns and generating reports that support strategic decision-making. This process requires having complete visibility of all information within a given scope while ensuring nothing is overlooked. Although algorithms and tools can assist with this analysis, security experts should be able to focus their attention on this critical analytical phase rather than on data collection and processing.

| *Challenges:* Classification.

By addressing the missing pieces and refining the methods used in the CTI cycle, organizations can better equip themselves to manage the increasing complexity and volume of cyber threats. Automating key processes, enhancing contextualization, and improving prioritization will help CERTs and SOCs stay ahead of potential attacks and strengthen their overall cybersecurity frameworks. Figure 3.1 depicts all steps of the CTI cycle and aligns all papers included in this work. My contributions focus on computation-based stages, whereas *Planning & Direction* and *Feedback & Evaluation* are structurally and politicly influenced stages.

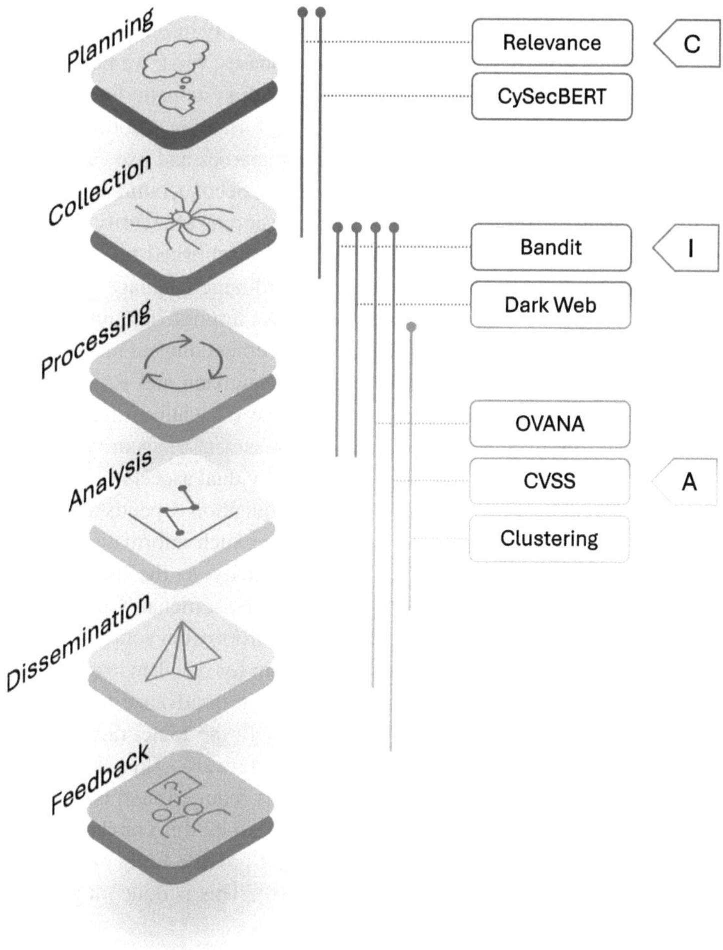

Fig. 3.1 Cyber Threat Intelligence cycle aligned with all papers and their domain CIA taxonomy (*conceptual*, *information*, and *analysis*). The paper abbreviations correspond as follows: *Relenvance* to Chap. 7, *CySecBERT* to Chap. 8, *Bandit* to Chap. 9, *Dark Web* to Chap. 10, *OVANA* to Chap. 11, *CVSS* to Chap. 12, and *Clustering* to Chap. 13

3.1.2 Research Approach

My research questions are directly aligned with the stages *Collection*, *Preprocessing*, and *Analysis* of the CTI cycle (*cf.* Fig. 2.2). **RQ1**[1] addresses the conceptual challenges inherent in managing diverse cybersecurity information sources and formats. Our methodology comprises three main phases: First, we develop a comprehensive notion of cybersecurity relevance through systematic literature review (*cf.* Chap. 7). This taxonomy serves to (i) classify information according to specific CTI types, (ii) map the landscape of information sources and use cases, and (iii) identify gaps and challenging information categories. Second, building on this taxonomical foundation, we collect and curate a diverse corpus of security-related texts, focusing on unstructured data that aligns with our established relevance criteria. This corpus serves as training data for a specialized large language model (LLM) (*cf.* Chap. 8), designed to enhance downstream tasks through domain-specific language understanding and more precise security text embeddings. Finally, we implement information quality (IQ) metrics based on Naumann and Rolker [97]'s IQ aspects (*cf.* Chap. 11), providing a comprehensive evaluation methodology for security IQ. This approach enables systematic assessment of security information sources, moving beyond isolated evaluations of individual aspects.

To answer research question **RQ2**[2], the final acquisition depends on the sources that are used. Usually one would use APIs to gather such information, if they are provided by centralized services. This comes in handy, as the final information usually follow a structured format like *json* and can be sufficiently parsed. But the focus of this work are unstructured, textual open information source, specifically, the World Wide Web (WWW). For this source, I aim for crawlers that gather information (*cf.* Chap. 9). As the topic is already known in advance (cybersecurity), I opt for a focussed crawler, that crawls in line with the topic, rather than gathering all information, which need to be filtered afterward. This approach keeps the technological footprint small (less bandwidth is required) and is able to adapt to different domains via the user-specified input. Despite this crawler prototype, I also take a closer look at the dark web, as it might provide further details, which cannot be found in the normal web (*cf.* Chap. 10). This is done on a qualitative

---

1 *"What is cybersecurity information, how is it represented, measured by quality, and grasped by Cyber Threat Intelligence use-cases?"*

2 *"What are efficient ways to automatically identify and find Cyber Threat Intelligence information in open sources?"*

base, as the deep web is even more dynamic than the normal web and implements further technological hindrances like Completely Automated Public Turing tests to tell Computers and Humans Apart (CAPTCHAs). The result is an assessment of openly available cybersecurity information found in the dark web, an overview of technological hindrances, and potential mitigations.

For research question **RQ3**[3] I want to first analyze the current state of information provided in open vulnerability databases. I exemplify this on the de facto industry standard vulnerability database National Vulnerability Database (NVD) and evaluate it with state-of-the-art methods (*cf.* Chap. 11), specifically Deep Learning (DL), machine learning (ML), and Natural Language Processing (NLP). Hereby, I aim to define an IQ metric to evaluate the NVD before and after applying the automated assessment methods. This is followed up by improvements to those methods and applying them to information of open source, specifically textual ones (*cf.* Chap. 12). The sources in this work are security advisories published by big tech-companies. We assess the severity of vulnerabilities based on the textual description alone. This allows a preemptive vulnerability prioritization before expert assessments are available [98–100]. Finally, the issue of information overload remains a significant challenge in the field of CTI. Therefore, I aim to examine the potential of DL and clustering in addressing this challenge (*cf.* Chap. 13). It focuses on methods that are accessible and straightforward, requiring minimal technical expertise, to ensure that CERT staff can readily integrate these techniques into their existing workflow without undue distraction.

The answers to all research questions contribute to answering the main research question *"How to leverage open sources for an automated and proactive Cyber Threat Intelligence"*. Table 3.1 provides an overview of the specific methods used for all papers in Part II. They range from reviews (*cf.* Chaps. 7 to 10), over different analysis of large cybersecurity datasets (*cf.* Chaps. 11 to 13), an implementation of a novel cybersecurity LLM (*cf.* Chap. 8), which supports other analysis in the security domain, to a prototyped focused crawler (*cf.* Chap. 9).

[3] *"How to analyze open information by unifying different inputs, removing overlapping information, and prioritizing the results to yield proactive Cyber Threat Intelligence?"*

Table 3.1 Overview of the methods used in the chapters of Part II

	Goal	Technique	Evaluation
7	Identifying a relevance notion for cybersecurity	Literature Review	–
8	BERT for CTI	Transfer-Learning with different CTI and security datasets	Domain-specific extrinsic, intrinsic, and general tasks
9	Efficient identification of new CTI information	Focused crawling with multi-armed bandit	Page relevance and harvest rate of crawler
10	Suitability of the dark web for automated information gathering	Qualitative analysis based on open dark web information sources	Manual evaluation based on strict methodology
11	Analysis of the IQ in the NVD	Classification, NLP	Classifier quality, inconsistencies in official source, missing information
12	Prediction of CVSS scores for open information sources	Downstream training of CVSS classifier	Classifier quality based on individual scores and overall score
13	Mitigate information overload through clustering	Requirement definition, selection of cluster algorithms and CTI datasets	Various clustering metrics, but mainly the homogeneity

3.2 Research Context

The present work is primarily located in ATHENE's research mission SecUrban but also contributed to the research project CYWARN. It uses the Lichtenberg High Performance Computer to train the necessary ML models and for evaluation.

The ATHENE mission SecUrban proposes means to enhance the cybersecurity and resilience of smart cities, through the proposal of security frameworks [101], research of cyber-arms control [14, 102], warning apps [9, 103], anomaly detection in urban infrastructures [104], and the support of communal CERTs [6]. It integrates data and information available in the web and smart cities, to enable stakeholders to more informed decision making. The main focus of my research is processes automation at CERTs and SOCs based on open, unstructured, and textual information [4–6]. However, I was also involved in research of warning apps [9] and cyber-arms control [14] as well.

The project CYWARN (13N15407) was supported by the German Federal Ministry for Education and Research (BMBF) and the present work provides additional content for this project between October 2020 and March 2024. It focused on supporting CERTs in their daily work with the analysis and preparation of cybersecurity information and the integration of it in a cybersecurity dashboard, with the aim to reduce information overload. Hereby our research groups focused on different aspects. First, the assessment of the daily work of CERTs and their current core problems [28], including prototyping interfaces for data-processing. Second, we dived deep into text augmentation to decrease the amount of necessary training data to cope with the omnipresent characteristic of crisis informatics by having no data-points before any crisis and either too much during the crisis or too few labeled data points and no capacity to label new data for, *e.g.*, classifier training tasks [105]. Third and last, we proposed methods to automate tedious and mundane tasks and thereby improve the work quality of security personnel in CERTs, which is the part, I was most involved in.

4 Findings

During the course of this work I tackled to implement methods to automate Cyber Threat Intelligence (CTI) processes, while including openly available information sources, most prominently, the World Wide Web (WWW). This advanced the field of CTI, but also the fields of Natural Language Processing (NLP) and Information Retrieval (IR). In the following, I present my findings aligned with my *CIA* classification of *conceptual*, *information* related, and *analysis* related research (*cf.* Table 1.1). This is followed up by a classification in the steps of the CTI cycle (*cf.* Fig. 2.2).

4.1 Conceptual, Information, and Analysis

My research can be categorized into three primary domains: *conceptual* research, *information*-related research, and *analysis*-related research. The conceptual research serves as a foundational framework that facilitates and enhances investigations in the other two domains. Information-related research focuses on advancing methodologies for information identification and acquisition, encompassing diverse approaches to gather data from multiple sources. Research in the analysis domain addresses methodological frameworks for information pre-processing, followed by systematic decomposition and extraction of actionable intelligence.

4.1.1 Conceptual

Before diving into the intelligence cycle there remains the necessity to provide an overview of what information security encompasses.

P. Kühn, *Proactive Cyber Threat Intelligence*, Technology, Peace and Security | Technologie, Frieden und Sicherheit,
https://doi.org/10.1007/978-3-658-49241-0_4

We provided an overview with Chap. 7 and outline a taxonomy of different types of security tools and their usage (*cf.* Fig. 7.1). This identified a total of six different sub-fields of security tools, of which two are in focus of the present work: *Intelligence* and *Vulnerability*. For those sub-fields the information available are standards and vulnerability databases, and different CTI sharing standards (*cf.* Table 7.1). The other sub-fields are mostly infrastructure specific, *e.g.*, configuration management tools, which notify of potential misconfigurations of services. Fig. 4.1 shows an

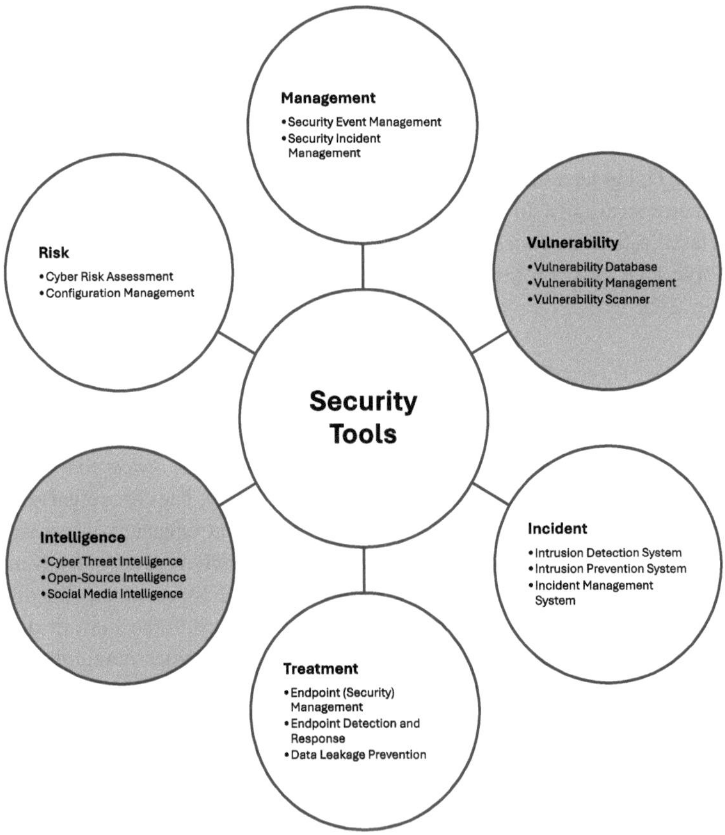

Fig. 4.1 Extract of the security taxonomy of Chap. 7. It shows six fields of security tools with their respective sub-fields *Management*, *Vulnerability*, *Incident*, *Treatment*, *Intelligence*, and *Risk*. The colored fields are in focus of the present work, while the uncolored ones are mainly infrastructure specific, *e.g.*, configuration management

excerpt of the taxonomy in Chap. 7 and highlights the sub-fields, that are relevant for this work.

The analysis step leverages machine learning (ML) and Deep Learning (DL), we opted for a customized cybersecurity large language model (LLM) to improve downstream tasks for the cybersecurity domain. Based on the above taxonomy, we compiled a natural language dataset composed of security-related social media short messages, reports, and papers, which is used to train this model. Table 4.1 provides an overview of the sources and number of tokens and entries composed in this dataset. We proposed CySecBERT, the first language model for cybersecurity, which is trained on a large, high-quality dataset with over 500M tokens, including security papers, vulnerability descriptions, blog posts, and social media short messages [2]. This model surpassed the state-of-the-art model for the security domain [36], while still mitigating catastrophic forgetting [106, 107]. We used this model in different papers in *collection* and *analysis* Chaps. 9 and 13. Combined with tools like Sentence-BERT (SBERT) [45] CySecBERT provides the ability to correctly embed incoming cybersecurity information[1]. Table 8.5 shows, CySecBERT is able to correctly infer terminology of the cybersecurity domain. This enables better filtering methods to exclude unnecessary information or correctly cluster them. Moreover, a correct inference of information allows for better classification or higher extraction precision of textual content (*cf.* Chaps. 9 and 12). This is especially important to mitigate information overload in the first place and for the analysis step afterward. Even semantically similar information, which either would not add to the overall picture can be detected this way to dismissed them, or if there are just singular information, that add new insights to fill missing gaps (*cf.* Chap. 13) (Table 4.2).

Table 4.1 Number of tokens for CySecBERT's dataset. The full table is depicted in Table 8.2

#Tokens	Sum	Mean	Entries
Blogs	169M	1.1k	151k
arXiv	167M	9.9k	16k
NVD	12M	71	171k
X (Twitter)	179M	45	4M
Total	528M	122	4.3M

[1] More general LLMs like BERT have difficulties to infer the difference of, *e.g.*, a *virus* in a medical or cybersecurity contexts.

Table 4.2 We evaluated CySecBERT through a comprehensive set of extrinsic and intrinsic tasks. The extrinsic evaluation utilized the full-shot CTI dataset from Bayer et al. [32] (F1), the CySecAlert dataset by Riebe et al. [11] (F1), and specific components from the OVANA dataset [5]: software versions (SV), software names (SN), and attack complexities (AC), measured by F1 scores. For intrinsic evaluation, we employed a cybersecurity word similarity task (F1) and a Log4j clustering task (silhouette score). Detailed experimental procedures and results are presented in Sect. 8.4. These experimental results were compiled and reported by [105]

Name	CTI	CySecAlert	SV	SN	AC	Word Sim.	Clu.
BERT	0.8599	0.8779	0.9247	0.8837	0.3323	0.4382	0.130
CyBERT [36]	0.8766	0.8647	0.9298	0.8834	0.3336	0.4861	0.113
CySecBERT	**0.8869**	**0.8883**	**0.9302**	**0.8871**	**0.3472**	**0.6382**	**0.172**

Furthermore, we provided an information quality (IQ) metric for vulnerability databases. As none such combined metric existed, we defined it on the criteria *comp* (*completeness*), *uniq* (*uniqueness*), and *acc* (*accuracy*) of the information associated with a single vulnerability information (*cf.* Sect. 11.3.3) [5]. Given a vulnerability v, *comp* measures, whether all information for v is available. *uniq* measures the amount of unique information of v, and *acc* measures the discrepancy of the provided information and the predicted information based on the vulnerability description. The composition of all three metrics make up our quality score $IQ = comp + uniq + acc$.

4.1.2 Information

The landscape of automated information collection can be broadly categorized into three approaches, each addressing different aspects of the collection challenge. *API-based collection* from official information sources represents the most stable approach but often comes with constraints such as rate limiting and incomplete coverage (*cf.* Chap. 11). *Domain-specific crawlers*, while offering precise data extraction capabilities, require substantial initial analysis and ongoing maintenance (*cf.* Chap. 12). These custom solutions become particularly problematic when websites undergo design changes, something we experienced during our work on Chap. 12. Their inherent specificity also limits their adaptability to new sources, making them suboptimal for expanding threat intelligence coverage. Meanwhile, *generalized crawlers* employing heuristic-based text extraction and offer broader coverage but may sacrifice precision (*cf.* Chap. 9). Table 4.3 summarizes the different crawling methods.

Table 4.3 Comparison of the different collection methods, their benefits, and caveats

Method	Benefits	Caveats
API-based	Structured output, stable, minimal maintenance	Rate limits, depends on offering party
Custom crawler	Structured output, no API necessary	High initial investment, prone to changes, high maintenance
General crawler	No limits, identification of new sources	High initial investment, unstructured output

Given that threat intelligence information is widely dispersed across the web, generalized methods have emerged as more promising than individual solutions in the context of the present work. Specifically, focused crawling in combination with a multi-armed bandit (MAB) has demonstrated particular effectiveness. I prototyped a focused crawler THREATCRAWL, which leverages three different methods to seek information: *forward search* (following links present on a page), *keyword search* (search for a page's keywords in search engines), and *backlink search* (find pages, that reference the given page). Its architecture is depicted in Fig. 4.2. The methodology's strength lies in its ability to identify and traverse relevant content through backlink analysis, which has proven especially effective in discovering previously unknown but highly relevant sources. When backlink traversal reaches saturation points, keyword-based searching serves as a complementary strategy for maintaining discovery momentum. Forward search is used to dive deeper into a given source and gather all available and relevant information. The results for a 2 000 steps run are presented in Table 4.4, which highlight the superiority of including backlink search in focused crawling. All Crawling methods, which include backlink search, either achieve the highest average similarity for the top method, the most crawled pages and new seed pages (pages with high similarity, excluding duplicates), or the highest harvest rate.

The dark web is an additional source of threat intelligence, though it comes with significant caveats. We did a qualitative analysis of 65 DWFs, 7 single-vendor shops, and 72 DWMs and searched them for openly available cybersecurity information. Thereby we concluded, that while the dark web may contain valuable information, the combination of social barriers (such as in-group behavior) and technical challenges (including widespread Completely Automated Public Turing test to tell

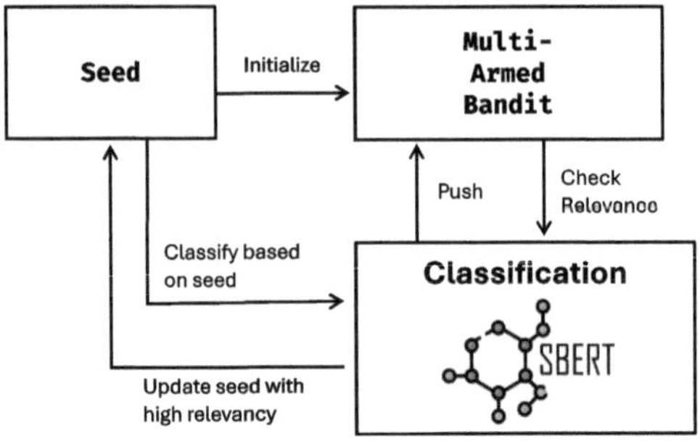

Fig. 4.2 Simplified architecture of the multi-armed bandit focused crawler. Given a seed, the prototype initializes the MAB with UCB-1 strategy, and follows up by using the UCB-1 recommendations afterward

Table 4.4 Results of THREATCRAWL evaluation (2 000 steps): Steps ($|S|$), total pages ($|P|$), relevant pages ($|P+|$), new seeds ($|Seed|$), harvest rate (HR [in %]), maximum similarity ($\lceil\sim\rceil$), domains ($|Dom|$), relevant domains ($|Dom+|$), top method (TM) with average similarity ($\overline{TM}$). Best and worst results highlighted

Method	S	$\lvert P\rvert$	$\lvert P+\rvert$	$\lvert Seed\rvert$	HR	$\lceil\sim\rceil$	$\lvert Dom\rvert$	$\lvert Dom+\rvert$	TM	$\overline{TM}$
TC_BFK		21 663	4 889	46	22.57	0.92	1 484	189	F	0.38
TC_BF		26 715	4 965	56	18.59	0.92	1 683	270	F	0.29
TC_FK		18 270	4 119	39	22.55	1	809	49	F	0.27
TC_BK		8 175	2 055	19	25.14	0.9	1 965	269	B	0.27
TC_F		21 710	4 992	48	22.99	1	372	52	F	0.27
TC_B	117	1 290	102	12	7.91	0.9	387	61	B	0.15
TC_K	22	155	5	1	3.23	0.81	111	5	K	0.24

Computers and Humans Apart (CAPTCHA) implementation and endpoint instability) makes reliable automated collection particularly challenging. Given these constraints, and the fact that surface web sources usually remain underutilized, prioritizing surface web collection optimization appears to be the more productive strategy for expanding threat intelligence coverage. If there are specific goals, which advocate the use of the dark web, *e.g.*, the surveillance of a certain dark web marketplace or a dark web hacker group, the dark web should of course be used. But if CTI surveillance is unsaturated by using the surface web, *i.e.*, the security team suffers from information overload by using information of the surface web alone, one should disregard the dark web for now, as it introduces potentially overwhelming maintenance steps.

These findings suggest that while diverse information collection methods each have their place in a comprehensive threat intelligence program, focused crawling of the surface web with emphasis on backlink analysis currently offers the most promising avenue for expanding intelligence gathering capabilities while maintaining reasonable operational overhead. This approach provides the best balance between discovery potential and implementation complexity in the current threat intelligence landscape.

4.1.3 Analysis

Initially, we analyzed the NVD in Chap. 11. Based on our research on the NVD, we were able to use Named-Entity Recognition (NER) extract missing information and propose updated severity scores with a weighted F1-score[2] of 0.886. A preliminary analysis showed gaps in vulnerability entry fields for which information is present in the vulnerability's description. This results in large clusters when searching for certain information (*cf.* Fig. 4.3). To address the challenges of incomplete vulnerability descriptions and missing data points, we implemented a NER approach based on the Bidirectional Encoder Representations from Transformers (BERT) model to identify potential Common Vulnerability Scoring System (CVSS) score information. We labeled a subset of our data according to CVSS attributes with dual objectives: (i) to calculate preliminary CVSS scores and (ii) to supplement missing CVSS scores. The classifier training utilized existing scores with their corresponding CVSS values. The NER models demonstrated varying performance levels, achieving F1 scores ranging from 0.999 for certain classes to 0.546 for others, indicating significant performance heterogeneity. The subsequent CVSS classifiers, which follow the NER stage, achieved scores up to 0.941 with other scores as low as 0.133. The high performance of NER models is particularly crucial given our pipeline's two-step approach, which relies on accurate keyword extraction during the initial NER phase. Lower performance metrics typically indicate insufficient training data starting in the NER stage passed on to the classifier (*cf.* Tables 11.2 and 11.3). This allowed us to analyze the free-form vulnerability descriptions for the whole NVD and identify several *batch added* entries (up to 117 at once) – entries which

[2] Accuracy $\in [0, 1]$ measures the proportion of correct predictions (both true positives and true negatives) among all predictions. Recall $\in [0, 1]$ (also known as sensitivity) represents the proportion of actual positive cases that were correctly identified. The F1 score $\in [0, 1]$ is the harmonic mean of precision and recall, providing a balanced measure that is particularly useful when dealing with imbalanced datasets.

are within a short timespan with no differentiating information. Other entries had outdated severity scores for which we calculated updated CVSS scores.

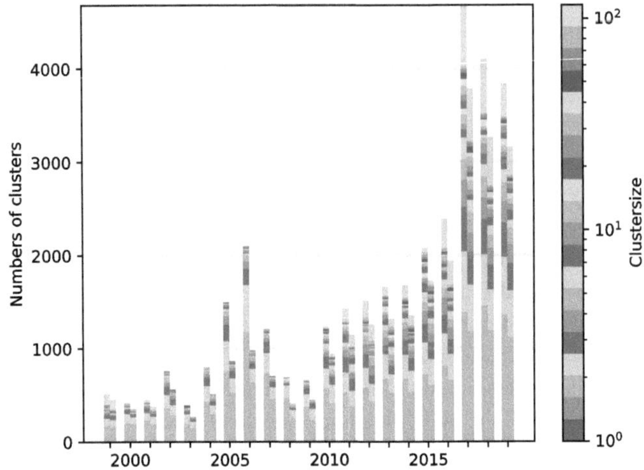

Fig. 4.3 Clusters by year excluding singletons: the first bar of each year shows the clusters of the original data and the second bar the clusters of the enriched data

This methodology is further refined in Chap. 12, where we analyze open information sources. We categorized top references into *version control system (VCS)*, *mailing list*, *patchnotes*, *security advisories*, *third party articles*, and *blog posts* (excerpt in Table 4.5). These categories were ranked based on uniqueness, structural uniformity, and natural language content(*cf.* Table 12.2). From these rankings, we selected various advisories, patch notes, and third-party publications for crawling and evaluation.

We evaluated multiple base models (BERT- SMALL, BERT- MEDIUM, and DISTILBERT), ultimately selecting DISTILBERT for optimization. Two variants were developed: one trained exclusively on expert-labeled data from NVD and Exploit-DB[3], and another incorporating all available ground truth data. Both models demonstrated comparable classifier quality and effectiveness in predicting CVSS scores for novel vulnerabilities. The models' compact size enables regular retraining to

[3] https://www.exploit-db.com/ is a database for publishing and explaining vulnerability exploits. We utilized exploit descriptions with known CVSS scores for our *high quality model*.

Table 4.5 Top referenced domains in NVD by group. #: rank, *Num.*: reference count, *Gr.*: assigned group, *Avail.*: domain accessibility status. Excerpt of Table 12.1

#	URL	Num.	Class	Avail.
1	github.com	25 064	1	✓
2	www.securityfocus.com	20 645	2	✓
3	www.securitytracker.com	10 842	2	✓
4	access.redhat.com	8 627	3/4	✓
5	support.apple.com	8 069	3	✓
6	lists.opensuse.org	7 930	2	✓
7	lists.fedoraproject.org	7 212	2	✓
8	www.oracle.com	7 006	3	✓
9	lists.apache.org	6 294	2	✓
10	www.debian.org	5 614	2/3	✓

address domain shift, and their potential integration with explainable AI could support vulnerability assessment by security personnel.

We investigated the challenge of information overload in Computer Emergency Response Team (CERT) environments and evaluated clustering methodologies as a mitigation strategy. Our comprehensive analysis of 196 distinct cluster-embedding combinations across five datasets revealed domain-specific performance characteristics. For security bug report (SBR) data, deep clustering methodologies combined with *stella* models achieved exceptional performance with homogeneity exceeding 0.88, while partition-based clustering paired with *NV-Embed-v2* demonstrated strong performance on advisory data (CySecAlert, homogeneity 0.61). The analysis of threat reports (ThreatReport) initially yielded suboptimal results (homogeneity 0.34), though subsequent parameter adjustments (increasing cluster count from 12 to 60) improved performance to exceed 0.5 homogeneity. All evaluated algorithms achieved dimensional reduction rates exceeding 90 %, effectively addressing the core challenge of information overload. Runtime performance proved highly efficient, with complete processing pipelines executing within approximately five minutes on consumer hardware, potentially saving over 3 750 work hours annually per analyst. However, persistent challenges were identified through auxiliary metrics, particularly near-zero silhouette coefficients indicating complex cluster structures across all datasets. Table 4.6 provides an excerpt of the results in Table 13.4. These findings suggest that while clustering approaches offer significant potential for information overload mitigation, their effectiveness varies by data type and may require careful parameter tuning for optimal performance in specific domains like threat report analysis.

Table 4.6 Clustering performance evaluation across five datasets, showing top-3 algorithm-embedding combinations per dataset. Results include ground truth (GT) cluster count (#C) and cardinality ($|c|$), along with achieved homogeneity (H), Silhouette coefficient (Sil), and predicted cluster count (#C). The THREATREPORT dataset was additionally evaluated with adjusted parameters (*n_clusters = 60*, denoted as THREATREPORT-60) to improve performance. Combinations are ranked (#) and sorted by homogeneity in descending order (↓) within each dataset category

	Algorithm	**Model**	**H ↓**	**Sil**	**#C**
	CYSECALERT (GT: $|c| = 13\,306$, #C = 2) — Use-case I				
1	K-Means	NV-Embed-v2	0.61	0.01	12.0
2	BruteForceK-Means	NV-Embed-v2	0.59	0.01	8.0
3	K-Means	stella_en_400M_v5	0.55	0.02	12.0
	MSE (GT: $|c| = 3001$, #C = 2) — Use-case I				
1	OPTICS	llama-3.2-1b	0.49	0.26	156.0
2	OPTICS	all-minilm-l12-v2	0.48	0.24	150.0
3	OPTICS	all-mpnet-base-v2	0.45	0.21	145.0
	THREATREPORT (GT: $|c| = 461$, #C = 39) — Use-case I				
1	DDC	gte-base-en-v15	0.34	0.02	30.4
2	DDC	all-minilm-l12-v2	0.33	−0.01	34.0
3	OPTICS	gte-large	0.32	0.08	23.0
	THREATREPORT-60 (GT: $|c| = 461$, #C = 39) — Use-case I				
1	AgglomerativeClustering	gte-base-en-v15	0.51	0.27	60
2	AgglomerativeClustering	all-mpnet-base-v2	0.50	0.27	60
3	AgglomerativeClustering	stella_en_400M_v5	0.50	0.28	60
	Security bug report (GT: $|c| = 5000$, #C = 5) — Use-case II				
1	DEC	stella_en_1.5b_v5	0.88	0.02	12.0
2	DKM	stella_en_1.5b_v5	0.87	−0.07	12.0
3	DeepECT	stella_en_1.5b_v5	0.86	−0.00	20.0
	SMS (GT: $|c| = 5574$, #C = 2) — Use-case III				
1	K-Means	NV-Embed-v2	0.89	0.02	12.0
2	DKM	NV-Embed-v2	0.88	0.01	12.0
3	BruteForceK-Means	NV-Embed-v2	0.88	0.03	8.0

4.2 Alignment with Cyber Threat Intelligence Cycle

This work advances multiple stages of the CTI cycle, particularly its computational aspects. In the *Collection* phase, we evaluated various approaches, demonstrating that focused crawling with backlink analysis offers the most effective method for expanding intelligence gathering while maintaining efficiency. The development of CySecBERT, trained on 500M+ tokens of security-specific content, enhanced not only the collection process through improved cybersecurity content identification but also subsequent phases.

For *Pre-processing*, OVANA established IQ metrics for vulnerability databases, focusing on *completeness*, *uniqueness*, and *accuracy*. This metric can be used to assess the IQ of vulnerability descriptions, and either filter information of bad quality or enhance it in further steps. CySecBERT facilitated accurate embedding of cybersecurity information, enabling enhanced information clustering. The *Analysis* phase benefited from high-quality NER models, allowing preliminary prioritization of critical vulnerabilities based on textual descriptions. Our clustering approach demonstrated significant potential for data point reduction, addressing information overload challenges. Analysis of the NVD revealed batch-added entries and outdated severity scores, emphasizing the importance of robust pre-processing to support expert analysis.

While this research concentrated on computational stages, it also supports the *Dissemination* phase through automated assessment tools and prioritization capabilities. The findings indicate that while automation can enhance all CTI cycle phases, human expertise remains essential for *Planning & Direction*, *Dissemination*, and *Evaluation & Feedback* stages. Table 4.7 summarize my contributions to improve the CTI cycle as a whole.

Table 4.7 Overview of my proposed improvements for the CTI cycle.

Phase	Improvements
Collection	Focused-Crawler
Pre-processing	IQ metric, improved embeddings
Analysis	CVSS prediction, clustering
Dissemination	CVSS values

Discussion

5

The previous chapter shows the applicability of open information sources in the context of the Cyber Threat Intelligence (CTI) cycle. In this chapter these findings are presented with regard to the research questions. Additionally, as the advent of foundational models is not part of the present work, but show tremendous potential in different aspects, I will discuss their limitations and potentials. This, and other limitations of my work, are part of Sect. 5.3. A more general outlook and future work are presented afterward.

5.1 Contributions

The presented work advances the field of CTI in various ways as seen in Chap. 4. All findings contribute to my overall research questions (*cf.* Sect. 1.2).

RQ1 **What is cybersecurity information, how is it represented, measured by quality, and grasped by Cyber Threat Intelligence use-cases?**

My research into the nature, representation, and measurement of cybersecurity information has yielded significant insights that advance our understanding of CTI information management. Through systematic analysis, we developed a comprehensive taxonomy of security tools that identified six distinct sub-fields, with our investigation focusing particularly on the *Intelligence* and *Vulnerability* domains as they relate to open-source CTI. The other fields are more focused on the actual detection of currently ongoing attacks using propretary information (*e.g.*, intrusion detection system). This taxonomical framework, illustrated in Fig. 4.1, provides a structured approach to understanding the diverse landscape of security tools and their applications.

P. Kühn, *Proactive Cyber Threat Intelligence*, Technology, Peace and Security | Technologie, Frieden und Sicherheit,
https://doi.org/10.1007/978-3-658-49241-0_5

A major contribution of our work was the development of CYSECBERT, the first specialized language model for cybersecurity. Trained on an extensive dataset of over 500M tokens encompassing security papers, vulnerability descriptions, blog posts, and social media content (Table 4.1), this model demonstrated superior performance compared to existing solutions. CYSECBERT particularly excelled in accurate embedding of cybersecurity terminology, enhanced filtering capabilities, improved classification and extraction precision, and better semantic understanding of security contexts. The model's effectiveness was validated through comprehensive evaluation across multiple tasks, as detailed in Table 4.2.

To address the fundamental challenge of information quality assessment, we introduced a novel information quality (IQ) metric for vulnerability databases. This metric is based on three key criteria: *comp* (measuring information completeness), *uniq* (quantifying unique information content), and *acc* (evaluating information accuracy against predictions). These components combine to form our IQ score through the formula $IQ = comp + uniq + acc$, providing a comprehensive framework for assessing vulnerability information quality. The integration of CYSECBERT with our IQ metrics created a robust system for processing cybersecurity information. The model's superior ability to contextualize cybersecurity terminology proved particularly valuable for mitigating information overload through intelligent filtering, identifying semantically similar information, detecting novel insights in seemingly redundant data, and supporting more accurate classification and extraction tasks. This capability is especially crucial given the scattered nature of CTI information across numerous sources.

Our findings demonstrate that while cybersecurity information remains distributed across diverse sources, the combination of specialized language models and quality metrics provides an effective framework for identifying, measuring, and utilizing CTI information. This approach not only enhances the accuracy of information processing but also addresses the fundamental challenges of information overload and resource allocation in cybersecurity operations. The empirical results suggest that our methodology offers a scalable and efficient solution for managing the complexity of modern cybersecurity information landscapes.

RQ2 **What are efficient ways to automatically identify and find Cyber Threat Intelligence information in open sources?**

Our investigation into efficient methods for automatic CTI information discovery and collection revealed several key insights. The research addressed the fundamental challenge of gathering scattered cybersecurity intelligence from across the web, where valuable information often remains undiscovered due to the manual nature of traditional collection methods.

Three distinct approaches to automated collection were evaluated: *API-based collection* (*cf.* Chap. 11), *custom crawlers* (*cf.* Chap. 12), and *generalized crawlers* (*cf.* Chap. 9). While API-based methods offer stability and structured output, they are constrained by rate limits and coverage. Domain-specific crawlers, despite their precision, proved problematic due to high maintenance requirements and vulnerability to website changes. The generalized approach, particularly focused crawling combined with multi-armed bandit (MAB), emerged as the most promising solution. Our prototype THREATCRAWL demonstrated particular effectiveness through its three-pronged strategy:

- Backlink analysis for discovering unknown but relevant sources
- Keyword-based searching for maintaining discovery momentum
- Forward search for deep exploration of identified sources

The empirical results showed that methods incorporating backlink search achieved superior performance across multiple metrics, including harvest rates up to 25.14 %, higher average similarities, and more effective seed page discovery. This suggests that backlink analysis provides an effective mechanism for expanding beyond known information sources while maintaining relevance.

Regarding the dark web as a potential source, our analysis revealed that while valuable information exists, the combination of social and technical barriers makes automated collection particularly challenging. Given that surface web sources remain underutilized, prioritizing their optimization appears more productive for general CTI gathering, though dark web monitoring may be warranted for specific intelligence objectives.

These findings suggest that focused crawling with backlink analysis offers the most promising approach for expanding intelligence gathering while maintaining reasonable operational overhead. This methodology effectively addresses the core challenge of automatically identifying and collecting relevant CTI information from open sources, providing a scalable solution that reduces manual effort while maintaining high information quality.

RQ3 **How to analyze open information by unifying different inputs, removing overlapping information, and prioritizing the results to yield proactive Cyber Threat Intelligence?**

Our investigation into proactive CTI through automated analysis and dissemination of security information yielded several significant methodological advances. At the core of our approach was the development of automated Common Vulnerability Scoring System (CVSS) score prediction from textual descriptions, addressing the predominantly text-based nature of security information on the web. Beginning with an analysis of the National Vulnerability Database (NVD), we developed sophisticated Named-Entity Recognition (NER) techniques that achieved a weighted F1-score of 0.886 for extracting missing information and proposing updated severity scores.

A key contribution was our cybersecurity-specific large language model (LLM), which enhanced downstream tasks including NER and vulnerability severity prediction. The NER models demonstrated varying effectiveness, with F1 scores ranging from 0.999 to 0.546 across different classes, while subsequent CVSS classifiers achieved scores up to 0.941. This methodology enabled us to identify several notable patterns in the NVD, including batch-added entries and outdated severity scores requiring updates.

To address information overload, we implemented clustering techniques that significantly reduced the number of individual data points while maintaining information integrity. We further refined our approach by analyzing diverse open information sources, categorizing them into distinct groups including *version control systems*, *mailing lists*, *patchnotes*, *security advisories*, *third-party articles*, and *blog posts*. Our evaluation led to the selection of DISTILBERT for optimization, with two variants developed: one trained exclusively on expert-labeled data and another incorporating all available ground truth data.

Importantly, our methods were deliberately designed to be independent of specific security standards like STIX or ATT&CK, following the KISS principle to focus on core methodologies rather than creating a comprehensive framework. This approach allows for flexible integration into existing tools through CTI feeds or direct implementation.

While our findings demonstrate the significant potential for automation in CTI analysis, they also underscore the necessity for careful implementation. Given that *every vulnerability is just a missed bug*, false negatives in automated systems could have catastrophic implications similar to unaddressed vulnerabilities. Therefore, these automated approaches should complement rather than replace expert oversight, particularly given the high-stakes nature of cybersecurity where even small false-positive rates can lead to critical oversights or resource misallocation.

The results conclusively demonstrate the viability of utilizing open information sources for proactive CTI. Through these automated approaches, we have shown that it is possible to significantly reduce the manual effort required for CTI analysis while maintaining high standards of accuracy and completeness in security assessments, particularly crucial given the rapid proliferation of open-source intelligence (OSINT) data and the constrained capacity of security personnel.

5.2 Improvements for the CTI Cycle

The research presented significant advancements across multiple stages of the CTI cycle, with particular emphasis on its computational aspects. In the *Collection* phase, focused crawling incorporating backlink analysis emerged as the most effective methodology for expanding intelligence gathering while maintaining operational efficiency. The development of CySecBERT, trained on over 500M tokens of security-specific content, enhanced both the collection process and subsequent phases through improved cybersecurity content identification and processing. For the *Pre-processing* phase, the introduction of the OVANA framework established crucial IQ metrics for vulnerability databases, focusing on three key dimensions: *completeness*, *uniqueness*, and *accuracy*. CySecBERT facilitated more accurate embedding of cybersecurity information, enabling enhanced information clustering and reducing information overload. The *Analysis* phase saw substantial improvements through high-quality NER models, which enabled preliminary prioritization of critical vulnerabilities based on textual descriptions. Our clustering approach demonstrated potential for significant data point reduction, directly addressing the challenge of information overload faced by security professionals. Analysis of the NVD revealed important insights, including the identification of batch-added entries and outdated severity scores, emphasizing the necessity for robust pre-processing to support expert analysis. While the research primarily concentrated on computational stages, it also enhanced the *Dissemination* phase through automated assessment tools and improved prioritization capabilities. These findings indicate that while automation can significantly enhance all phases of the CTI cycle, human expertise remains essential for the *Planning & Direction*, *Dissemination*, and *Evaluation & Feedback* stages (Table 5.1).

Table 5.1 Summary of contributions to the CTI cycle improvement

Phase	Key Contributions
Collection	Focused crawler with backlink analysis, CySecBERT model
Pre-processing	OVANA IQ metrics, enhanced embeddings
Analysis	NER models, clustering, CVSS prediction
Dissemination	Automated assessment tools, prioritization frameworks

5.3 Limitations

The present work focused on Natural Language Processing (NLP) models that could be run locally and privately, aligning with the requirements of security personnel who cannot share infrastructure details with third parties. However, the advent of foundational models like GPT, Gemini, or Claude presents both challenges and opportunities to this approach. While these models require significant computational resources, often necessitating external compute farms, they offer substantial potential for enhancing various aspects of our work.

Foundational models could significantly improve our current results, which achieved F1 scores of 0.941 for CVSS prediction and reduced data points by more than 90 %. Their capacity to generalize across tasks and domains could enhance web crawling by returning structured outputs without predefined parsers, potentially improving our current harvest rates. Preliminary analysis already demonstrates substantial improvements in performance metrics, even in zero-shot settings [105]. The context length capabilities alone are remarkable – Google's Gemini 1.5 Pro, for instance, allows context lengths of 2 million tokens [108], equivalent to $\sim$ 1.3 million words, compared to Sentence-BERT (SBERT)'s maximum of 384 tokens [45].

The emergence of smaller, locally-deployable models like *Llama3.2 3B* suggests a potential middle ground between privacy requirements and advanced capabilities. Such models could enhance our current clustering and classification approaches while maintaining data privacy. Additionally, they could enable more sophisticated recommendation systems that suggest prioritization based on vulnerability severities, current threat trends, and infrastructure-specific contexts, potentially improving upon our current classification accuracy.

This tension intensifies with the current proposal of open-source foundational model DeepSeek R1 [109], which outperforms current proprietary models like OpenAI o1. The fact that it is open-sources enables it to be run locally and, as such, privacy preserving, while still providing the outstanding performance of foundational models.

A significant limitation of our work is the focus on computational methods without direct incorporation of security personnel feedback. While we utilized established metrics from NLP and Information Retrieval (IR) domains and considered requirements identified in previous studies [27, 28], future work should include direct validation with security professionals. This could involve comparative studies of manual versus automated analysis times, assessment of false positive/negative rates in operational settings, and evaluation of our prioritization methods' effectiveness in real security operations.

Several promising research directions remain unexplored. Social graph analysis of threat actors and security information could provide additional CTI by combining tactic, technique, and procedure (TTP)s and enriching actor profiles. Log file analysis, while infrastructure-specific, could enhance our current methods through anomaly detection and pattern recognition. While we deliberately excluded social media analysis due to its extensive coverage in existing research [11, 34] and current data access challenges [110], future work could explore integration points with our methods as social media platforms evolve.

Both cybersecurity and crisis informatics heavily rely on the effective collection, processing, and analysis of information from diverse sources to support decision-making processes. The improvements to the CTI cycle, particularly through focused crawling, enhanced pre-processing with IQ metrics, and advanced analysis capabilities using NER models and clustering techniques, mirror similar advancements in crisis informatics where information overload and rapid response requirements pose comparable challenges [111, 112]. Just as crisis informatics leverages social media and public information sources for situational awareness, the enhanced CTI cycle demonstrates the value of incorporating diverse information sources through automated collection and analysis, while maintaining human expertise in critical decision-making phases. This parallel evolution in both fields emphasizes the growing importance of combining automated information processing with human expertise to address complex, time-sensitive challenges in both cybersecurity and crisis response scenarios.

Regarding CTI feeds, Bouwman et al. [95] demonstrated limited overlap and significant timeliness issues among existing solutions. While our work focused on underlying information processing methods rather than direct comparisons with CTI feeds, future research could explore how our approaches might improve existing feed quality and timeliness. This is particularly relevant for national Computer Emergency Response Team (CERT)s, some of which are restricted from using commercial tools.

Integration with existing CTI platforms represents another important direction. While we deliberately focused on independent methods rather than full-fledged frameworks, future work could explore how our approaches complement and enhance current commercial and open-source solutions. This could include developing standardized interfaces for our methods to integrate with existing platforms while maintaining their independence and flexibility.

The rapid advancement of foundation models also suggests the potential for new approaches to information synthesis and analysis. Future research could explore how these models might enhance our current methods for vulnerability assessment, information clustering, and threat prioritization while maintaining the privacy and security requirements of the cybersecurity domain. This includes investigating hybrid approaches that combine our local, specialized models with the capabilities of larger foundation models in privacy-preserving ways.

5.4 Outlook and Future Challenges

Simon [113] correctly identified in 1971 that there is no scarcity of information but one of attention. This observation has become increasingly relevant over the decades, particularly with the advent and spread of the World Wide Web (WWW). The progression from manually indexed directories like *Yahoo* and *Aliweb* in the early 1990s to automated solutions like *The Lone Wanderer*[114] and ultimately Google's *PageRank*[115] highlights both the overwhelming volume of available information and the necessity for automated solutions to address information overload challenges.

Our research contributions directly address these challenges in the cybersecurity domain. Our CySecBERT model and automated CVSS prediction methods, achieving F1 scores of 0.941, demonstrate that effective security analysis can be accomplished without relying on complex, proprietary solutions. Similarly, the evaluated cluster-embedding combinations reduced data points by more than 90 % while maintaining cluster homogeneity, which shows that complexity reduction and effective security aren't mutually exclusive.

However, the cybersecurity industry faces a crucial paradox: solutions developed to address information overload often introduce new complexities and vulnerabilities – the classical Rebound effect [116]. This is particularly evident in recent developments where security vendors themselves have become potential attack vectors. As Greenberg [117] noted: "For years, it's been an inconvenient truth within the cybersecurity industry that the network security devices sold to protect customers from spies and cybercriminals are, themselves, often the machines those intruders hack to gain access to their targets." Recent incidents involving major security vendors [118–122] further underscore this concern.

The situation has become so problematic that even GCHQ employees have compared current security practices to "medieval witchcraft" [123]. Our research suggests a different path forward. The demonstrated effectiveness of our open-source approaches, combined with our IQ metrics for evaluating information quality, offers a more transparent and verifiable approach to security analysis. Rather than adding more layers of proprietary tools, organizations could benefit from streamlined, automated analysis pipelines that prioritize transparency and verifiability.

Looking forward, the emergence of foundation models presents both opportunities and challenges. Our research suggests a hybrid approach might be optimal—combining local processing for sensitive data with the enhanced capabilities of larger models. This could build upon our current methods while addressing privacy concerns. The key will be maintaining the transparency and accountability demonstrated in our research while leveraging these advanced capabilities.

Future work must address several critical areas:

- Integration of our clustering and classification methods with existing security architectures while maintaining simplicity
- Development of standardized interfaces for our methods to complement rather than replace existing solutions
- Expansion of our IQ metrics to evaluate not just information quality but also the effectiveness of security tools themselves
- Investigation of hybrid approaches that combine local and foundation models while preserving privacy and security requirements

The success of future developments will depend on whether the industry can break free from the cycle of introducing new vulnerabilities while attempting to fix existing ones. Our research demonstrates that effective security analysis doesn't require ever-increasing complexity. Instead, thoughtfully designed automated systems, transparent evaluation metrics, and open-source approaches could provide a path toward more robust and manageable security solutions.

The evolution from early web directories to current security challenges demonstrates that adding more tools isn't always the answer. Future solutions must balance automation and complexity reduction while maintaining robust security measures. Our research provides a foundation for this transformation, showing how automated analysis, quality metrics, and information clustering can contribute to more streamlined, thoughtful, and inherently secure systems.

Conclusion 6

This dissertation has investigated the challenges and opportunities in leveraging open-source information for automated and proactive Cyber Threat Intelligence (CTI). Through systematic analysis and development of novel methodologies, we have demonstrated significant advances in automating critical aspects of the CTI cycle while maintaining high standards of accuracy and reliability.

My contributions span three primary areas. First, we developed a comprehensive taxonomy of security tools and information sources, providing a structured framework for understanding the diverse landscape of CTI resources. Second, we created CySecBERT, a specialized language model for cybersecurity that achieved superior performance in domain-specific tasks, demonstrating F1 scores up to 0.941 in Common Vulnerability Scoring System (CVSS) prediction. Third, we evaluated novel clustering and embedding combinations, reducing data points by more than 90 % while maintaining cluster homogeneity.

The development of a focused crawler, ThreatCrawl, with its innovative combination of forward search, keyword search, and backlink analysis, demonstrated the feasibility of efficient automated intelligence gathering. Our information quality (IQ) metrics for vulnerability databases established new standards for evaluating information quality, providing a foundation for more reliable threat intelligence assessment.

These achievements, however, must be viewed within the broader context of cybersecurity challenges. As my research shows, the increasing complexity of security solutions often introduces new vulnerabilities—a rebound effect that threatens to undermine the very protections these systems aim to provide. Our findings suggest that effective security analysis doesn't necessarily require increasingly complex solutions, but rather thoughtfully designed automated systems that balance efficiency with transparency.

P. Kühn, *Proactive Cyber Threat Intelligence*, Technology, Peace and Security | Technologie, Frieden und Sicherheit,
https://doi.org/10.1007/978-3-658-49241-0_6

Looking forward, the emergence of foundation models presents new opportunities for enhancing CTI capabilities, though their implementation must be carefully considered to maintain privacy and security requirements. The future of CTI likely lies in hybrid approaches that combine the privacy advantages of local processing with the enhanced capabilities of larger models, while maintaining the transparency and accountability demonstrated in our research.

Looking forward, the emergence of foundation models presents both opportunities and challenges for enhancing CTI capabilities. Their implementation must be carefully considered to maintain privacy and security requirements. The future of CTI likely lies in hybrid approaches that combine the privacy advantages of local processing with the enhanced capabilities of larger models, while maintaining the transparency and accountability demonstrated in our research.

This work sets the foundation for future research in automated CTI analysis. While our methods have shown promising results, their true value lies in demonstrating that automation can effectively support, rather than replace, human expertise in cybersecurity. As the field continues to evolve, the principles of transparency, efficiency, and careful validation established in this research will become increasingly important for developing robust and reliable security solutions.

Future work in the realm of CTI will involve a comprehensive approach, encompassing all phases of the CTI lifecycle: Direction, Collection, Processing, Analysis, Dissemination, and Feedback. Building upon foundational methods like specialized language models (*e.g.*, CySecBERT) and advanced clustering techniques that efficiently manage vast amounts of data, future efforts will focus on seamlessly integrating these analytical capabilities with existing security architectures. This includes developing standardized interfaces to ensure these methods complement rather than replace current solutions. Furthermore, expanding the evaluation metrics to not only assess information quality but also the effectiveness of the security tools themselves will be critical for continuous improvement. Finally, a key area of investigation will be hybrid approaches that combine the privacy benefits of local processing for sensitive data with the enhanced capabilities of larger foundation models, all while rigorously preserving privacy and security requirements throughout the entire CTI pipeline. This holistic approach aims to provide transparent, verifiable, and actionable intelligence to diverse stakeholders, from strategic decision-makers to operational security teams, ultimately moving the industry towards more robust and manageable security solutions that break the cycle of introducing new vulnerabilities.

The success of future developments will depend on whether the industry can break free from the cycle of introducing new vulnerabilities while attempting to fix existing ones. The challenges of information overload and cybersecurity will continue to grow, but this research demonstrates that thoughtfully designed automated systems, combined with rigorous quality metrics and open-source approaches, can provide a path toward more manageable and effective security solutions. The future of CTI lies not in adding more layers of complexity, but in developing smarter, more efficient ways to process and analyze the wealth of available information while maintaining the highest standards of security and reliability.

Part II
Publications

7 The Notion of Relevance in Cybersecurity: A Categorization of Security Tools and Deduction of Relevance Notions

Abstract

Proper cybersecurity requires timely information to defend the IT infrastructure. In a dynamic field like cybersecurity, gathering up-to-date information is usually a manual, time-consuming, and exhaustive task. Automatic and usable approaches are supposed to be a solution to this problem, but for this, they require a notion of information relevance to distinguish relevant from irrelevant information. First, on the basis of a literature review, this paper proposes a novel cybersecurity tool categorization based on corresponding tool types with their respective definitions and core features. Second, it elaborates information used in each category and deduces notions of relevance. Third, it outlines how these findings informed the design of a security dashboard to guide Computer Emergency Response Team staff in identifying current threats in open-source intelligence sources while mitigating information overload.

Original Publication ***Kuehn, P.**, Bäumler, J., Kaufhold, M.-A., Wendelborn, M., Reuter, C., "The Notion of Relevance in Cybersecurity: A Categorization of Security Tools and Deduction of Relevance Notions," in* Workshop-Proceedings Mensch und Computer, *Darmstadt: Gesellschaft für Informatik, 2022.* https://doi.org/10.18420/muc2022-mci-ws01-220.

Minor changes regarding language, style, and formatting have been made in this chapter to provide a consistent experience.

P. Kühn, *Proactive Cyber Threat Intelligence*, Technology, Peace and Security | Technologie, Frieden und Sicherheit,
https://doi.org/10.1007/978-3-658-49241-0_7

7.1 Introduction

Staying up-to-date with cybersecurity (CySec) information is crucial to secure one's system. It is one of the tasks Security Operations Center (SOC)s and Computer Emergency Response Team (CERT)s are doing on a daily basis. Newly found vulnerabilities, which remain undetected in one's IT infrastructure, can be the source for tremendous damage [124]. While traditional security checks like penetration testing and system hardening are good investments to identify known vulnerabilities, approaches like Cyber Threat Intelligence (CTI) are necessary complements for a proper cyber defense [33]. CTI facilitates information sharing on incoming threats in partner organizations and automates the processes of information gathering and correlation from open-source intelligence (OSINT) sources. This requires reliable sources of information on cyber threats, which, however, are usually identified manually [27].

Automated tools for gathering CTI have already been developed, but they either use information sources known beforehand (*e.g.*, security advisories or blogs) [33] or are limited to Twitter [125]. Although these sources might be relevant for the specific use-cases, other scenarios in cybersecurity (CySec) require different sources. A systematic identification of relevant information, in known and unknown sources, is not possible, because, to the best of our knowledge, there is currently no systematically identified notion of relevance. This leaves the domain of CTI, with a focus on OSINT sources, to experts who either use known information sources or have sufficient experience in the field to quantify the relevance of sources themselves.

Furthermore, in order to achieve usability and the successful adoption of tools, they often must be tailorable according to the individual or organizational needs, preferences, or definitions of actionable or relevant information [126–129]. This paper aims to close these gaps by (i) providing an overview of different CySec categories and corresponding types of tools, (ii) deducing relevance notions for each tool category, and subsequently (iii) presenting their application for an automated and usable CySec dashboard.

Thus, we seek to answer the following research questions.

RQ1 **What different categories of CySec tools are used in research and practice?**

RQ2 **Which security related relevance notions are found in these categories?**

7.2 Related Work

7.2.1 Relevant Information for Cyber Situational Awareness

In order to prevent and respond to attacks effectively, CySec staff needs to establish situational awareness and undertake informed decisions [130]. Endsley [131] coined a distinction of three phases to attain situational awareness: the perception of the elements, the comprehension of their meaning, and the projection of their status. Cyber Situational Awareness (CSA) refers to a state of knowledge of actors that enables them to perceive the relevant elements in the cyber environment within a certain volume of time and space, to comprehend their meaning, and project their status in the near future [132]. Although it was established as a subset of situational awareness, it cannot be considered in isolation because events in cyberspace usually impact the physical world, *e.g.*, financially, socially, or politically. Its growing importance within public administration has been reflected in numerous national CySec strategies [133]. Especially professionals in formalized security organizations such as CERTs, whose work is particularly challenging due to the necessity of coordination and information exchange within and beyond the team, benefit from a continuous improvement of their situational awareness [134].

The term CSA is often related to knowledge about occurrences in one's own network, but CySec staff have to look way further [130]. This includes information about current threats events, new vulnerabilities, possible mitigations, and new technologies. However, due to the steadily increasing number of attacks and security breaches, the volume and variety of potentially relevant data and data sources is steadily increasing. Thus, obtaining and ensuring adequate CSA is increasingly dependent on the properties and capacities of the tools in use [135]. Husák et al. [132] identify both the joint aggregation, analysis, and visualization of data from as diverse sources as possible in real-time, as well as the support of operators in assessing the veracity and credibility of the provided information as novel challenges that have so far been addressed only very fragmentary in technology development. Even recently developed dashboards to support CSA focus on data from within the organization, such internal network security data and vulnerability management [136], while external data sources are rarely included and credibility assessment of social media data is not supported [12].

Though several notions of relevant information for CTI are discussed in literature [33, 35], there is a lack of a systematic investigation of these notions based on a comprehensive overview of tools in the field of CySec rather than on insights into individual aspects.

7.2.2 Data Sources & Tools for Cyber Threat Intelligence

At present, several data sources are available to obtain relevant information and enhance CSA. Skopik et al. [137] identified vulnerability databases, specialized search engines, and alerting systems as key data sources for cyber incident response. One of the most well-known vulnerability databases is the *National Vulnerability Database (NVD)* [138], but other well-known alternatives are the commercial tool *Vulners*[1] or the open-source database *Open CTI*[2]. In these databases, the Common Vulnerabilties and Exposures (CVE) have become established as the standard. All publicly disclosed security vulnerabilities in software or hardware components are cataloged there, supplemented by an assessment of their criticality in the form of the Common Vulnerability Scoring System (CVSS) score. This score takes into account various factors, *e.g.*, the complexity for the attacker to exploit the vulnerability. However, vulnerability databases should not be used as the only source for attaining CSA, since they contain only publicly disclosed vulnerabilities [125], entries may be inconsistent [5], and vulnerabilities are not the only threats [33]. Furthermore, many manufacturers publish information about vulnerabilities and fixes on their own channels first [139] by means of security advisories.

Besides vulnerability databases and security advisories, CTI platforms have been established to allow both the collection and analysis of cyber threat data [140]. To support this task, several tools, like the Malpedia database [141], exist. Indicator of Compromise (IoC)s are also often used to improve attack prevention, and tools such as MISP [142], ThreatFox [143], and Pulsedive [144] were designed to allow the collection and processing of IoCs. In recent years, social media has become another important data source for CTI. CySec experts can exchange threat information more quickly on those platforms than by using more formal channels [145]. A lot of social media intelligence (SOCMINT) tools have been developed to facilitate the collection and analysis of data from social media, also in other domains [146]. Regarding cyber incident response, Mittal et al. [145] created a Twitter-based warning framework for CySec incidents, and Rodriguez and Okamura [147] developed a CSA system aimed at retrieving and classifying security relevant information from the same platform. Still, the available tools are not designed to support CySec staff through the whole process of multi-platform data acquisition, data analysis, and cyber threat communication [12].

---

[1] https://vulners.com/

[2] https://www.opencti.io/en/

Although we identified several data sources and tools for gathering CTI, we were unable to find a comprehensive categorization of CySec tools, which will be subject of the next section before a deduction of relevance notions follows.

7.3 Categorization of Cybersecurity Tools

In this section, we aim to give an overview over the different CySec tools used in research and practice, their respective use-cases, and their superordinate categories.

7.3.1 Method

First, in order to design a categorization of CySec tools for professionals, web searches were performed to get an overview of available categories, tools, and already conducted tool comparisons. A particular emphasis, in addition to established taxonomies, was put on systematic surveys and market studies of tools of one type in scientific publications, on websites for CySec practitioners, and on websites of CySec software vendors. Based on this, a list of relevant tool types and exemplary systems was derived and discussed among the authors. Our team of authors consisted of two researchers in human-computer interaction and CySec, respectively, one researcher and graduate student in CySec, and one graduate student in political science, all of whom were involved in the interdisciplinary research project CYWARN [10]. We excluded tools for non-professional end-users (*e.g.*, internet security suites) and grouped tools with very similar purposes under a uniform designation. It should be emphasized that the delineated tool types represent ideal types.

Second, after mutual agreement on the scope of considered tools, another search was conducted. Google Scholar was used to collect scientific literature on the characteristics, core features, and definitions of the previously identified tools, resulting in 16 relevant references for analysis. We selected Google Scholar as search engine, since it indexes publications from numerous computer science relevant databases, *e.g.*, IEEE-Xplore, ACM Digital Library, and USENIX. While no systematic literature review was conducted, the premise of the search and selection was to identify scientific and, if possible, peer-reviewed fundamental literature on tool types, which contains a concise tool definition. If multiple publications for one tool type were found, it was checked whether the definitions referred to different core features, and if this is not the case, the most cited publication was selected. The main rationale driving our high-level categorization was to assign various tool types to one of six key operational scenarios in CySec practice. Tool types that could not be appropriately allocated to one of the six thematic categories were grouped into the seventh category, *others*. Besides the descriptive (*cf.* Sect. 7.3.2) and visual presentation (*cf.* Fig. 7.1) of our categorization, we provide an overview of the tool definitions and core features found in scientific literature.

7.3.2 Cybersecurity Tool Hierarchy

Our categorization of CySec tools comprises the seven categories of *(a) management*, *(b) vulnerability*, *(c) incident*, *(d) treatment*, *(e) intelligence*, *(f) risk*, and *(g) others*.

The category of *management* comprises tools for Security Event Management and Security Information Management, which are often combined as Security Information and Event Management. A Security Event Management tool is used for monitoring, processing, and correlation of log and event data to enable incident response [148]. Thus, it primarily assists analysts during the detection and real-time analysis of security-relevant information. A Security Information Management tool, on the other hand, uses log data from local systems and applications to primarily support compliance reporting, internal threat management, and monitoring [148]. Often it allows for automatic reconciliation of log data with pre-set organizational standards.

The second category of tools is centered around hardware and software *vulnerabilities*, including vulnerability databases, vulnerability management tools, and vulnerability scanners. Vulnerability databases are platforms for the collection, dissemination, and maintenance of information about known security weaknesses in both computer system hardware and software [149]. Besides the vulnerabilities themselves, assessments of their severity and in-depth background information are often provided. Vulnerability management tools identify, assess, and fix vulnerabilities to minimize risks [150]. They typically incorporate the functionality to categorize, prioritize, and analyze vulnerabilities, are interoperable with vulnerability scanners, and support during patch management. Finally, vulnerability scanners scan for security vulnerabilities in networks or host systems [151].

Another category of tools is designed around CySec *incidents*, including Intrusion Detection Systems, intrusion prevention systems, as well as incident management software. are designed to support the detection and isolation of intrusions in computer systems [152]. . intrusion prevention systemss feature an extended range of functions as they, *i.e.*, in addition to the features of an Intrusion Detection System have capabilities to prevent intrusions [153]. Incident management tools, in contrast, primarily support the structuring and planning of workflows after the detection of security incidents, *e.g.*, through the collection of incident data and communication within the organization [154].

The fourth category is related to the *treatment* of cyberattacks, which comprises tools of endpoint (security) management, endpoint detection and response, as well as data loss/leakage prevention systems. Central endpoint (security) management tools enable inventory management, network access, and asset control of computer equipment of an organization as well as the monitoring and patching of installed software on endpoints, thus facilitating compliance with software license agreements, regulatory standards, and internal policies [155]. In contrast, endpoint detection and response systems monitor endpoints for suspicious behavior [156]. Beyond monitoring security-related activities, the tools also provide guidance for the reaction to and removal of cyber threats on endpoints [156]. Data loss/leakage prevention systems tools aim to detect and prevent potential data breaches in a timely manner by monitoring data while in use on endpoints, in motion as part of network traffic, or during storage [157].

The following category includes *intelligence* tools which are differentiated either as CTI, OSINT, or SOCMINT tools. CTI tools and providers provide low level IoCs or high level threat actor behavior information from different sources [8, 43]. One common feature is the extraction of IoCs from various data sources. In the context of CySec, the use of OSINT involves monitoring openly accessible websites, forums, and marketplaces in clear- and dark web to collect CTI [158]. While OSINT can be extracted from a wide variety of different sources, SOCMINT tools use social media [159, p. 205] for the acquisition of CTI. Both, advanced OSINT and SOCMINT tools, may include information prioritization and credibility assessment functionalities.

Another category is related to the assessment of CySec *risks*, including cyber risk assessment and configuration management tools. Cyber risk assessment tools execute risk-assessment frameworks, provide automated solutions for infrastructure operators to determine the present state of their CySec efforts and, if required, identify specific areas for improvement [160]. A different type of tool with relevance for risk assessment has emerged to support configuration management [161]. These tools inform operators of misconfigured systems according to specific guidelines.

Finally, the category *others* comprises CySec tools, *e.g.*, packet sniffers, sandboxes, firewalls, deception tools, or penetration testing tools, which fit in none of these categories. While these tool types serve important functions within CySec operations, they are not described in detail in this section. Thus, they have also been omitted in Fig 7.1.

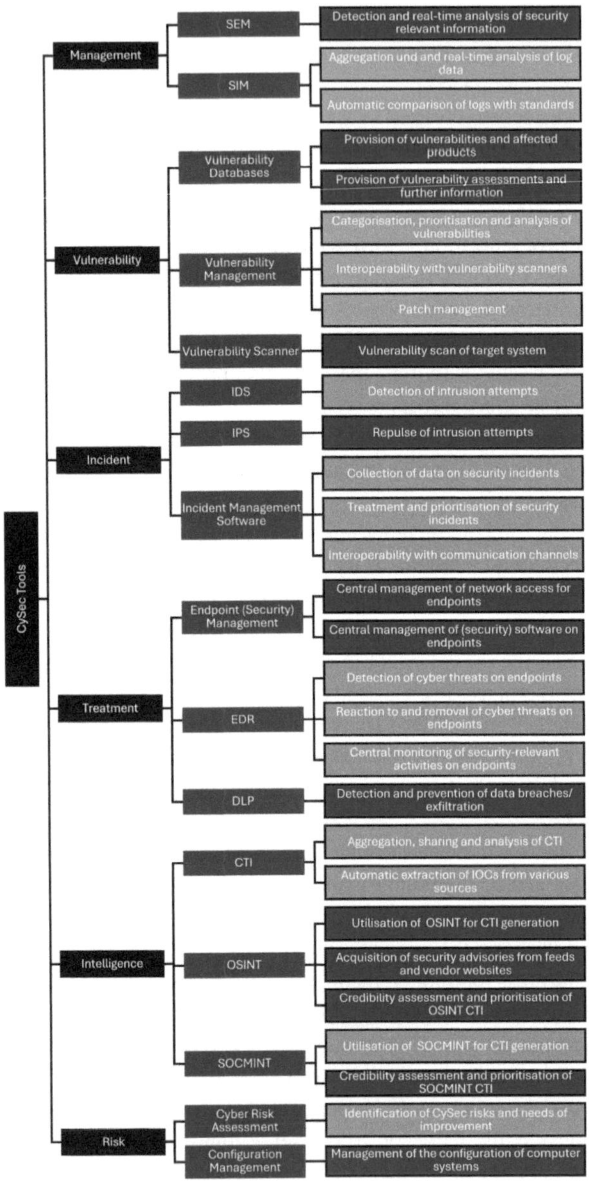

Fig. 7.1 The diagram shows a hierarchical overview of categories of cybersecurity tools. Each category is divided into several subcategories

7.3.3 Implications for Notions of Relevance

To answer *RQ1*, we searched for existing CySec tools and tool taxonomies, analyzed 16 academic publications to identify definitions and core functionalities of the tools, and finally derived a novel high-level categorization on this basis (*cf.* Sect. 7.3.2).

Since this is an ideal-typical categorization and the actual range of features of tools available on the market varies greatly, some tools can be clearly assigned to one tool type and category, e.g., cyber risk assessment and data loss/leakage prevention systems, while others combine different functionalities and can thus be assigned to several tool types and categories at once. For instance, Security Information and Event Management (SIEM) tools that combine the features of Security Information Management (SIM) and Security Event Management (SEM) are now predominant [92], and some CTI systems provide OSINT and SOCMINT in addition to intelligence from commercial or closed sources. Likewise, some tools offer interfaces for integrating data from other tool types; e.g., vulnerability management tools partly feature interoperability with vulnerability scanners as well as CTI platforms [162], and some SIEM systems also provide CTI [163]. Furthermore, the tool types may contribute to the different aspects of CSA; network, threat, and mission awareness [164]. While SIM, configuration management, endpoint (security) management, and vulnerability management tools, as well as vulnerability scanners, may primarily contribute to network awareness, tools of the intelligence and incident categories as well as SEM, data loss/leakage prevention systems, and endpoint detection and response tools along with vulnerability databases may particularly improve threat awareness. By contrast, cyber risk assessment tools may support the establishment of mission awareness.

The diverse operational scenarios of the tools within CySec, the broad spectrum of information types processed and generated, and the contribution of the tools to different aspects of CSA complicate the derivation of a comprehensive notion of relevance for CySec as a whole. Such an universally applicable notion would remain at a very abstract level and would be of limited use in identifying relevant sources and information for specific operational scenarios and tools. Instead, it seems reasonable to derive separate notions of security relevant information for each of the categories identified in this chapter (*cf.* Sect. 7.4).

7.4 Deduction of Relevance Notions

Relevance is a subjective information criteria, but necessary to mitigate or even prevent information overload. While information might be relevant for one user, it might be irrelevant for another [111, 165, 166]. Hence, different notions of relevance for the CySec domain based on the identified tool categories are necessary (*cf.* Sect. 7.3). Those categories focus varying information types and notions of relevant information.

To gain an overview over the information used in these categories, we analyze related work that propose methods for (automatic) analysis of information in the corresponding operational scenario. These results can be leveraged in subsequent research to guide automatic analysis methods to improve one's CySec. Hereby, we split the relevance notion into *context* and *precise* notions. An *context relevance notion* indicates the scope of information (*e.g.*, infrastructure), while *precise relevance notions* refer to information of a specific type (*e.g.*, blog posts or source code).

The *management* category, comprised of Security Event Management-, Security Information Management-, and Security Information and Event Management-tools, focuses on the local infrastructure. All information surrounding the infrastructure in scope is relevant, *e.g.*, log files, traffic information, and infrastructure information (soft-, hardware). Accordingly, Debnath et al. [35] propose a tool for anomaly detection in IT infrastructures, using unsupervised learning to detect patterns in application log files. Another system, proposed by Moh et al. [167], detects web attacks by applying a multi-stage log file analysis based on combination of pattern matching and supervised machine learning techniques.

The *vulnerability* category leverages different weakness information sources. First, information in vulnerability databases is primarily used [5, 38, 39]. Precise information that is used are Common Platform Enumeration (CPE), Common Weakness Enumeration (CWE), CVSS, Common Attack Pattern Enumerations and Classifications (CAPEC), and Open Vulnerability and Assessment Language (OVAL) [168]. Second, source code is another primary information source. You et al. [169] leverage source code, vulnerability descriptions, and issue descriptions to nail down vulnerabilities for automatic exploit creation.

The *incident* category uses network-based information. Jyothsna et al. [170] summarize recent studies in the field of anomaly based intrusion detection and explain why signature based anomaly detection is preferred for mainstream implementations. Signature based detection systems use either context information, *i.e.*, metadata retrieved during attacks, or content information, *i.e.*, payloads, to filter malicious content. Common standards combine these information, *e.g.*, SNORT or

YARA. Ou et al. [171] implement a host-based Intrusion Detection System (IDS) that combines log file analysis technology with back-propagation techniques of neural networks to improve the efficiency and accuracy of IDS. Kumar and Sangwan [172] introduce a signature based IDS based on SNORT.

The *treatment* of threats use data at rest (inactive data), data in motion (any network data), or data in use (*e.g.*, RAM or cache data) [173]. One use case is the identification of the attack vector. These systems leverage information which are similar to that presented in the incident category, *e.g.*, network traffic, log files, or file hashes, domain names, and byte sequences (known as IoCs). A discussion of various possible attack scenarios shows that most state-of-the-art endpoint detection and response systems, which analyze system logs, fail to prevent the bulk of the attacks [107]. Alneyadi et al. [174] provide a survey on data loss/leakage prevention systems showing that such systems are based on firewalls, Intrusion Detection Systems, and virtual private networks.

Intelligence tools use different sources containing varying types of information from the OSINT domain. Some systems [33, 34, 57] crawl grey literature like websites, blogs, or social media to identify current threat information. They focus on IoCs and CVE information. Others use solely social media to identify intelligence information in order to obtain an overview of currently used vulnerabilities, *e.g.*, CVE, CWE, CPE information, and tactic, technique, and procedure (TTP) [125, 175, 176]. There are different standards to publish IoCs, namely the STIX-, OpenIOC-, and MISP-standard [33, 57].

Risk assessment in CySec rests upon internal and external information. Lyu et al. [177] assess safety and security risks of cyber-physical systems. The assessment considers asset values, potential threats, and vulnerabilities. It is based on sector-precise standards [178, 179], the used infrastructure, and uses related management-, vulnerability-, and intelligence-information (*cf.* the previous paragraphs). Aksu et al. [180] introduce a CySec risk assessment methodology that is asset- as well as vulnerability-centric. . Kure et al. [181] introduce a CySec risk assessment methodology that considers risk from a holistic stakeholder model perspective and integrates the cascading effects from interdependent cyber-physical system components. The information used for this approach are information about vulnerabilities, threats, and risks to an asset.

Tools unlisted in other categories have cyber threats (*cf.* vulnerability and intelligence) as a common baseline. All of them need information about vulnerabilities and threats (*e.g.*, CVSS-scores, signatures, and exploits) as well as information about the systems they protect (*e.g.*, system logs, operating system, and network ports).

Building on the categorization and findings of Sect. 7.3, we examined related work on automatic information analysis methods across different operational scenarios of CySec. Based on this, we elaborated context and precise notions of relevant information (*cf.* Table 7.1) as well as key information standards in the different CySec categories, thus answering *RQ2*. There are a few base categories, *e.g.*, management or vulnerability, whose information are used in others, *e.g.*, risk. In their current form, these notions remain vague due to the numerous categories in CySec, with the most precise information being the named standards. The information used in CySec is basically every stream of information in or connected to ones infrastructure (*e.g.*, system-, application-, and network-log files) as well as related external information and standards listed in Table 7.1.

Table 7.1 Summary of context and precise information in each category. Italic information relate to other categories

Category	Context	Precise
Management	Infrastructure	Log files, used hard- and software-information (CPE)
Vulnerability	System weakness	Vulnerability databases, source code, CVSS, CPE, CWE, CAPEC, OVAL
Incident	Infrastructure, network, metadata	Log files, network traffic, payloads, SNORT, YARA
Treatment	Attacker behavior, metadata	Log files, network traffic, IoCs
Intelligence	Blogs, threat information, *vulnerabilities*	IoCs, CVSS, CWE, CVE, attacker behavior
Risk	*Mangement, vulnerability, intelligence*	CPE, CVE, CVSS

7.5 Towards an Automated and Usable Cybersecurity Dashboard

In order to capture relevant information for CySec intelligence and vulnerability management in an mostly automatic and usable manner, we employ a human-centered process to iteratively design and refine the Open Data Observatory, a CySec dashboard to visualize relevant CySec information across multiple channels. We shortly introduce the methodology of user evaluation, followed by the presentation of the dashboard.

7.5.1 Method

The requirements for our first iteration of design were identified through interviews with people working at or with CERTs in two rounds in 2019 and 2021. To acquire the necessary data, requests for semi-structured expert interviews were sent out in two rounds. After receiving the participants' acceptance and informed consent, each interview session lasted around 50 minutes using a web conferencing tool. In the first round of interviews (n = 8), a stronger emphasis was put on organizational factors and collaborative practices. The interview guideline comprised nine open-ended questions organized in three parts: (i) an introduction of the interviewee and his/her organizational role, (ii) the deployment, organization, and work processes of the CERT, and (iii) the communication and cooperation between CERTs.

To get further insights into technology use by CERTs, we conducted a second round of interviews (n = 7). In the second round, the perspective of some non-CERT organizations that share information with CERTs has also been included. The second interview guideline comprised technology-focused questions on the (i) interviewees' role and organization, (ii) reporting of cyber incidents, (iii) monitoring of cyber incident data (*e.g.*, IoCs), (iv) analysis, prioritization, and verification of gathered evidence, as well as (v) communication of recommendations and warnings. We interviewed seven cyber incident managers, five team leaders, two information security officers, and one public safety answering point for CySec issues.

With the consent of the participants, all interview sessions were recorded and later transcribed. In order to account for the rich, qualitative interview data, we decided to conduct an inductive qualitative content analysis where categories emerge from the data analyzed [182]. Thus, the first and second authors independently employed open coding [183] to gather data into approximate categories reflecting the issues raised by respondents based on repeated readings of the data and its organization into similar statements. The authors then compared their categories and compiled a list of the most important user requirements identified during the coding process. Based on this, all authors participated in a workshop to identify the requirements guiding the implementation of the Open Data Observatory.

The identified user requirements that serve as the foundation of our design and development process are displayed in Table 7.2. While the next subsection will briefly present details on the design of the CySec dashboard, a first evaluation of the CySec dashboard using scenario-based walkthroughs and corresponding design implications can be found in a different study [184].

Table 7.2 Overview of the derived user requirements

#	User Requirements
1	Manual efforts of data collection should be replaced by (semi-) **automation**.
2	Enable **modularity** to add new data sources and features in the later course of development.
3	Allow gathering of data from different sources and unify collected data for **interoperability**.
4	Offer ways to support **data protection** (*e.g.*, anonymization and data sparsity).
5	Automatically detect and filter out **redundant** information across different sources.
6	Allow the **visualization** of important data to get an overview and accelerate decision making.
7	Facilitate information **management** for different users or organizational roles.
8	Allow **customization** of data sources, filters, features, and settings to fit individual needs.
9	Display only **priority** (relevant) information to prevent the overload of human capacities.
10	Evaluate information based on trustworthiness and provide data to the user for **verification**.

7.5.2 Cybersecurity Dashboard

From an architectural point of view, the Open Data Observatory is a web application based on Vue.js as the overall framework, Bootstrap for responsive design, and Chart.js for data visualization. Besides some local filtering options, all other actions of the Open Data Observatory, such as searching for posts in open and social media or managing users, are forwarded to a backend called Open Data API. This API is implemented following the paradigm of a web-based and service-oriented architecture (SOA). It is a Java Tomcat application using the Jersey Framework for RESTful web services and the MongoDB database for document-oriented data management. Several libraries facilitate the automated and continuous real-time collection of data from open sources, such as NVD vulnerabilities, IoCs, and RSS feeds, or social media source APIs for Flickr, Reddit, Tumblr, Twitter, and YouTube.

The interface (*cf.* Fig. 7.2) comprises up to four feeds with security advisories, CVEs, IoCs, and social media data to fully cover the intelligence and vulnerability parts of our idenfied CySec category (*cf.* Sect. 7.3). The security advisories are embedded via RSS feeds provided by software and hardware vendors, and the API of the NVD database is used to populate the CVE feed with documented vulnerabilities. Furthermore, we decided to use the ThreatFox platform to obtain IoCs, and individual platform APIs to gather information from social media (*e.g.*, Reddit or Twitter). For each feed, specific charts and a different set of available and characteristic information per entry is displayed (*e.g.*, a textual description and the CVSS score for CVEs). The displayed data set can be selected in the upper left corner and is based on predefined individual parameter settings that are used to query the various sources. On demand, users can display or hide individual or all feed entries and pin important entries that are then highlighted and displayed in the black bottom pin menu for quick access. In order to filter the displayed information, users can either click on the interactive charts (*e.g.*, on the critical bar to show only critical vulnerabilities within the CVE feed) or open an advanced filtering menu with additional options (*e.g.*, also filtering by the CVE ID or the affected product of a vulnerability). In the top-right corner, a full-text search field allows searching for keywords across the different feeds simultaneously. To render the collection and processing of data in the application transparent, each displayed information contains a link to the external source (*e.g.*, the NVD for the vulnerabilities).

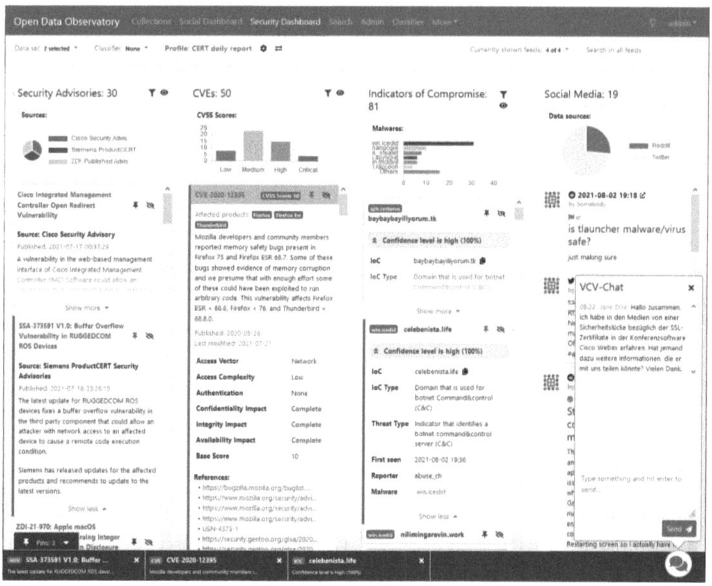

Fig. 7.2 Interface of the Open Data Observatory, a cross-platform cybersecurity dashboard featuring security advisories, common vulnerabilities and exposures, Indicators of Compromise, and social media feeds

7.6 Conclusion

CySec is one major cornerstone to secure organizations and corporations, especially due to the increasing digitization. However, proper CySec has many different facets, depending on the specific operational scenarios and the applied tools, and requires to be up-to-date with information of relevance, *e.g.*, on vulnerabilities and attack campaigns. This information is usually gathered from the web or domain specific sources and assists security experts in their task. Yet, there exists no study on what types of information is relevant for different tool categories in the field of CySec.

This paper answers which categories of CySec tools are used in research and practice *(RQ1)*, by presenting six categories and their features in detail (*cf.* Sect. 7.3), and which security related relevance notions are found in these categories *(RQ2)*, outlining information contexts and specific information types (*cf.* Sect. 7.4). These results then inform the implementation of a security dashboard to guide CERT staff in their task of identifying current CySec information in OSINT sources, which mitigates the problem of information overload. We already conducted a first evaluation of the presented tool [184] and its findings will be used to refine the interface for a better (*e.g.*, by the inclusion of machine learning classifiers for relevancy assessment [185]) and more usable detection of relevant CySec information. Our results present a solid baseline to identify new information of relevance, but should be further enhanced as part of a thorough study into each tool category.

CySecBERT: A Domain-Adapted Language Model for the Cybersecurity Domain

8

Abstract

The field of cybersecurity is evolving fast. Security professionals are in need of intelligence on past, current and—ideally—on upcoming threats, because attacks are becoming more advanced and are increasingly targeting larger and more complex systems. Since the processing and analysis of such large amounts of information cannot be addressed manually, cybersecurity experts rely on machine learning techniques. In the textual domain, pre-trained language models like BERT have proven to be helpful as they provide a good baseline for further fine-tuning. However, due to the domain-knowledge and the many technical terms in cybersecurity, general language models might miss the gist of textual information. For this reason, we create a high-quality dataset (https://github.com/PEASEC/cybersecurity_dataset) and present a language model (https://huggingface.co/markusbayer/CySecBERT) specifically tailored to the cybersecurity domain which can serve as a basic building block for cybersecurity systems. The model is compared on 15 tasks: Domain-dependent extrinsic tasks for measuring the performance on specific problems, intrinsic tasks for measuring the performance of the internal representations of the model as well as general tasks from the SuperGLUE benchmark. The results of the intrinsic tasks show that our model improves the internal representation space of domain words compared to the other models. The extrinsic, domain-dependent tasks, consisting of sequence tagging and classification, show that the model performs best in

---

Minor changes regarding language, style, and formatting have been made in this chapter to provide a consistent experience.

P. Kühn, *Proactive Cyber Threat Intelligence*, Technology, Peace and Security | Technologie, Frieden und Sicherheit,
https://doi.org/10.1007/978-3-658-49241-0_8

cybersecurity scenarios. In addition, we pay special attention to the choice of hyperparameters against catastrophic forgetting, as pre-trained models tend to forget the original knowledge during further training.

Original Publication *Bayer, M.,* ***Kuehn, P.,*** *Shanehsaz, R., Reuter, C., "CySecBERT: A Domain-Adapted Language Model for the Cybersecurity Domain,"* ACM Transactions on Privacy and Security, *Mar. 15, 2024.* https://doi.org/10.1145/3652594.

| CORE A[1] with IF 3.0[2]

8.1 Introduction

Cybersecurity has become an increasingly critical concern in today's digital world, as the number of cyber attacks continues to rise and their consequences become more severe [186]. The COVID-19 pandamic[3], the Russian war of aggression against Ukraine [187] and the resulting shifts in diplomatic and geopolitical dynamics have only intensified this development. As a result, there is a growing need for effective cybersecurity measures to prevent, detect, and respond to these threats. One of the key components of effective cybersecurity is the ability to analyze and understand the vast amounts of data generated by various sources, such as logs, network traffic, and threat intelligence reports. However, much of their analysis is still performed manually by security experts, which is a time-consuming and labor-intensive process.

To address these challenges, recent advancements in Natural Language Processing (NLP) and machine learning have shown promise in automating the analysis of cybersecurity-related data. For instance, Cyber Threat Intelligence (CTI) [27], an important domain of cybersecurity, involves the collection and analysis of information about emerging threats and vulnerabilities. Information thereby is often disseminated in the form of Indicator of Compromises (IoCs) or unstructured natural language text in blog posts and news articles [188, 189]. Applying NLP techniques to CTI can help automate the extraction and understanding of relevant, evidence-based knowledge, thus significantly reducing the manual workload for experts [190].

[1] Based on the latest CORE journal ranking (2020).

[2] Updated 5. Februar 2025.

[3] https://enterprise.verizon.com/en-gb/resources/articles/analyzing-covid-19-data-breach-landscape/

With regard to underlying NLP mechanisms, word embedding methods that use a sparse vector space to represent words are prominent examples [191]. In this context, models such as Bidirectional Encoder Representations from Transformers (BERT) [47] have become the standard basis models in all machine learning tasks that involve natural language as input. These models are already pre-trained on a general level and can be adapted to the task at hand by so-called fine-tuning. However, research has shown that the full potential of such models cannot be realized when applied to domain-specific tasks [48, 192–194]. This is intuitive because these models intend to cover as many domains as possible and, especially in normal-sized models, specific domain knowledge is lost due to capacity constraints or because the knowledge is not even included in the training data. To gain domain-specific knowledge, pre-trained models can be further trained on domain-specific corpora to achieve better results in this particular domain [48]. However, it must be ensured that catastrophic forgetting does not occur, which means that the model forgets its original knowledge.

Models trained on general domain corpora, such as Wikipedia, and without further training often reach their limits when applied to domain-specific tasks such as cybersecurity [195]. Their limitations can be explained by two primary reasons:

1. Unfamiliarity with domain-specific terminology: General domain models may not have encountered specific vocabulary of the cybersecurity domain, such as names and designations of new vulnerabilities or unique threat actor groups. This lack of exposure can lead to reduced performance when analyzing cybersecurity texts, as the model may fail to recognize crucial information.
2. Semantic ambiguity across domains: General domain models may fail to disambiguate words with different meanings in different contexts. For example, the word *virus* might be interpreted by a general model as referring to a biological disease rather than to a type of malware, which is the more relevant meaning in the context of cybersecurity [195].

In this paper, we propose CYSECBERT, a word embedding model based on BERT [47] for analyzing cybersecurity texts. Our aim is to enable state-of-the-art NLP for the security domain and to provide a model which is highly suitable for practical cybersecurity use cases and a solid base for further research in this field. By evaluating our resulting model on different tasks (*i.e.*, intrinsic and extrinsic tasks) we ensure that it indeed enriches the cybersecuritydomain. In a preliminary evaluation phase, we try to identify appropriate hyperparameters to minimize the problem of forgetting previously trained knowledge, which is then verified using a standard NLP benchmark. In this study, we pre-train a model on a thoroughly chosen cybersecurity corpus consisting of various datasets, such as scientific papers, Twitter, webpages, and the National Vulnerability Database. A well-performing model for this use case may supersede high manual workload for researchers and experts. Although there are well performing models for various very specific purposes in the cybersecurity domain [54, 196], the importance of a general cybersecurity model that can serve as a basis for all kinds of tasks is undeniable. The following contributions are made by this paper:

Contributions

C1 A pre-trained, general purpose cybersecurity language model based on BERT, called CySecBERT.
C2 Experiments for hyperparameter tuning in light of catastrophic forgetting.
C3 An evaluation of CySecBERT based on several tasks tailored to the cybersecurity domain, including intrinsic and extrinsic tasks, as well as general benchmark, to measure if and to which degree the model forgets past knowledge.
C4 A comparison of our model to a related cybersecurity model and to the original Bidirectional Encoder Representations from Transformers (BERT) model, as well as a discussion about its shortcomings and potential improvements.

8.2 Related Work

This subsection provides an overview of relevant work on the topic of BERT models. We thereby outline models adapted to different domains that have emerged following the publication of BERT. Moreover, we summarise work that already proposes BERT-like language models for the cybersecurity domain. Finally, we specify the research gap that we intend to fill in the scope of our research.

8.2.1 BERT Models in Different Domains

In various publications, researchers have demonstrated that it is possible to achieve good classification quality on domain-specific text corpora with pre-trained models such as BERT. Of particular interest for our research is the method of domain-adaptive pre-training (DAPT) [192], which describes the process of training an already pre-trained language model on a domain-specific, domain-dependent dataset and which is conducted in the same way as the pre-training. Hence, this technique differs from classical fine-tuning in that the model is not specialised for just one task, but serves as a building block for many tasks in the field. It has been implemented in several other domains since the introduction of BERT [36, 48, 193]. A prominent example is BIOBERT, introduced by Lee et al. [48], where BERT was adapted to a biomedical corpus. BIOBERT was initialized with weights from Devlin et al. [47]'s BERT model and then pre-trained once again, this time with a large biomedical dataset, where the dataset was more than five times larger than BERT's. Evaluating the resulting model with a subsequent fine-tuning process of three different biomedical text mining tasks, which are Named-Entity Recognition (NER), Relation Extraction (RE), and Question Answering (QA), Lee et al. [48] were able to largely outperform BERT and previous state-of-the-art models on these aforementioned tasks. Models that address other domains present similar approaches. SCIBERT [193] for example focuses on scientific publications whereas DA- ROBERTA, introduced by Krieger et al. [197] covers media bias. Gururangan et al. [192] underpin our method of additional pre-training on BERT by yielding good results in their application of this approach on ROBERTA [198], a variant of BERT which uses the same transformer-based architecture. In contrast to studies such as BIOBERT by Lee et al.[48], in which only a single domain at a time is considered, Gururangan et al. [192] covered several different domains.

Similarly, researchers have also explored BERT models for the cybersecurity domain. For example, Ranade et al. [36] propose a BERT model for this domain called CyBERT. Although their paper states that fine-tuning on BERT took place, in fact, they further pre-trained BERT for the cybersecurity domain. Fine-tuning is performed on top of this pre-trained cybersecurity model and is primarily used for application. In general, however, their research goal is similar to ours.

Moreover, there are further cybersecurity BERT models, which, however, are fine-tuned instead of subjected to continued pre-training, as is required for true domain-adaptive pre-training (DAPT). Hence, this makes them less suitable for other task of the cybersecurity domain. MALBERT [54] is a BERT-based model from the cybersecurity domain focusing on the detection of malicious software. Another security-related work is CATBERT, introduced by Lee et al. [196]. They replaced some transformer blocks with adapters and fine-tuned the BERT model for the detection of phishing emails. Mendsaikhan et al. [199] introduced a BERT-based Natural Language filter for identifying and classifying cyber threat-related information from publicly available information sources with high accuracy.

An overview of the approaches with their domains and how they are trained can be found in Table 8.1. These works are related to ours because the approach of adapting BERT to a specific domain is similar to our methodological framework and differs mainly in the specific target domain. All in all, the different pre-trained BERT approaches can be of use for our work as an orientation and for comparisons of our results to theirs w.r.t. performance. Notwithstanding the fact that BERT has achieved great results in various domains, the full potential for the cybersecurity domain has yet to be exploited.

Table 8.1 Overview of relevant BERT models for special domains. The method explains if the model was only fine-tuned (FT) or also pre-trained (PT)

Model / Paper	Domain / Use Case	Method	Model Base
BioBERT [48]	Biomedical	PT (+ FT)	BERT
SciBERT [193]	Scientific	PT (+ FT)	BERT
[192]	Papers (bio. + CS), news, reviews	PT (+ FT)	RoBERTa
MalBERT [54]	Malware	FT	BERT, RoBERTa, DistilBERT
CatBERT [196]	Phishing	FT	DistilBERT
ExBERT [200]	Exploit prediction	FT	BERT
[199]	CTI	FT	BERT
CyBERT [36]	Cybersecurity	PT	BERT

8.2.2 Research Gap

The previously established research gap has led us to develop a model with the aim of achieving satisfactory performance for cybersecurity textual material in various tasks. BERT has already been transferred to different domains, resulting in domain-specific models (BioBERT [48], SciBERT [193]) and has been applied even to specific domains in the cybersecurity field, producing models like MalBERT [54] or CatBERT [196]. Besides, there are also BERT models such as from Mendsaikhan et al. [199] that are not specific to a particular cybersecurity domain and that can handle various different texts, but are then fine-tuned for a particular task, as in the case of the BERT model from Mendsaikhan et al. [199] CTI classification. In contrast to these models and associated development research, the proposed CySecBERT is able to handle many different sources and text forms, and is at the same time the basic building block for all cybersecurity-related tasks.

As outlined in the introduction, there are a multitude of research problems in the field of cybersecurity, based on the essential aspect of Information Extraction. A solid method to address this can improve research in the cybersecurity field at a stroke. Furthermore, the outcome enables extensibility and the application of additional layers on top of the model, for instance CRF [201], (Bi)LSTM, or both combined [202].

Ranade et al. [36] also address the delineated research gap to some extent. Unfortunately, no juxtaposition was made with the results of BERT as the baseline, but only a presentation of their model's outcome was given. We compare our CySecBERT with theirs, which is varied in the model training and the corpus [36]. Furthermore, in delimitation to their work, we evaluate a whole span of different cybersecurity tasks, ranging from classification to NER and clustering tasks and we include the results of BERT for comparison. We also take into account the phenomenon of catastrophic forgetting, where the pre-trained model forgets its already acquired knowledge in the new training phase. This issue has not been targeted in other works of this area. The similarity in in the work of Ranade et al. [36] and our work results from the nature of the research task, which all the more underlines the importance of the approach. We understand that the attention given to this research gap is important and encouraging at the same time. Multiple works addressing a similar objective can be complementary and thus accelerate filling the gap in research. Nonetheless, our work distinguishes itself from the work of Ranade et al. [36] at several points including the evaluation step, the applied data, and overall by the extent of our work.

8.3 Methodology

This section provides a brief background on domain adaptive pre-training, including the planned training process, presents the dataset which is used to adapt our proposed language model to the cybersecurity domain, the architecture of the model, and a pre-evaluation phase.

8.3.1 Domain Adaptive Pre-Training

DAPT of language models to a specific domain is a common method to achieve an advanced domain-specific language model (*cf.* Sect. 8.2). It has shown to increase model performance in several ways. Not only is model performance on downstream tasks increased, hence generating better evaluation results, but training time is also reduced for such tasks due to smaller datasets for the training process to achieve similar performance. These prospects lead us to expect that the cybersecurity domain will benefit greatly from a domain-adapted, pre-trained language model for every possible task, *e.g.*, NER and relevance classification, to name a few.

We aim to adapt BERT to the cybersecurity domain based on domain specific text corpora (*cf.* Sect. 8.3.2) [192]. Our DAPT pipeline is built with *Huggingface*[4] and *Weights and Biases*[5]. The final domain-adapted pre-trained model is based on `bert-base-uncased`. Likewise, the text corpus is tokenized using the `bert-base-uncased` model. The training itself is done on the Lichtenberg Cluster[6].

During the training phase, we try to mitigate the problem of catastrophic forgetting [106] by reducing the learning rate, the training steps, and the size of the dataset compared to BERT pre-training. In this way, the susceptibility to catastrophic forgetting should be greatly reduced because the new learning process is subordinated. Nevertheless, we test whether the problem also occurs with the created model by evaluating it on a non-cybersecurity task. While we expect no improvements, we want to analyze to what extent the old knowledge is altered.

8.3.2 Text Corpus

When creating the text corpus, we paid great attention to the quality of the data, as this quality transfers to the model [203]. The text corpus is composed of four different sub-corpora: (i) blog data, (ii) arXiv data, (iii) National Vulnerability Database (NVD) data, and (iv) Twitter data. Our decision for this selection is based on the kind of information that is used by security professionals, and on the fact that these sources are commonly used in recent publications regarding machine learning. Furthermore, those sub-corpora vary considerably in their structure. While the NVD contains short, objective and precise language with semi-structured information, and Twitter consists of short posts with objective, subjective, emotional, on- as well as off-topic, etc. content. arXiv encompasses long papers with highly educational language, and blog posts are typically longer articles with less formal language.

The blog posts build a solid foundation for different writing styles and practical information in information security, including vulnerability and exploit information, threat notifications [33], and foundational knowledge. We initially aimed for a set of

---

[4] https://huggingface.co/

[5] https://wandb.ai/

[6] http://www.hhlr.tu-darmstadt.de/

41 different domains to be crawled. During the automated crawling process, three of them either blocked our requests, or contained so many advertisements and cookie information that we decided to omit those sources. This resulted in 38 different blogs which we crawled, like troyhunt.com, darkreading.com, schneier.com, and krebsonsecurity.com, to name a few. Besides, we filtered duplicates and instances shorter than 300 characters[7] and filtered for English texts using fasttext [204]. This process resulted in a total of over 151k blog posts, based on the initial list of 165k web pages.

Secondly, we use arXiv papers from the category *Cryptography and Security*[8] [48]. Due to errors during the text extraction process, we ignored papers with lower length than 3000 characters, resulting in over 16k papers.

Thirdly, we use vulnerability descriptions of the National Vulnerability Database (NVD) [5, 39]. Experts curate those texts[9], which is why they need no further processing. Hence, we do neither filter nor pre-process these information.

Finally, we use Twitter as information source [11, 125, 175], where we crawled datasets containing the following keywords based on information we received from Computer Emergency Response Team (CERT) members:

Keywords
infosec OR *security* OR *threat* OR *vulnerability* OR *cyber* OR *cybersec* OR *infrasec* OR *netsec* OR *hacking* OR *siem* OR *soc* OR *offsec* OR *osint* OR *bug-bounty*

Additionally, we crawled dedicated datasets of data breaches, like the Microsoft Exchange Server Data Breach, for example. Of all the tweets collected, we only used those that were annotated as English by Twitter. While the posts include retweets, we did not gather replies. Overall, we managed to crawl nearly 4M tweets with over 179M tokens in total.

We would like to emphasize that all collected entries have been minimally edited at most, as we want to have the most natural form in which the model will later function. A summary of all datasets is depicted in Table 8.2. To get an overview

[7] A randomly selected and manually inspected sample of corner case blog posts, *e.g.*, shorter or longer posts, showed that short posts contained mostly advertisements or cookie notifications.

[8] For text extraction we used opendetex for papers in tex format or textract for papers in pdf format.

[9] https://www.cve.org/ResourcesSupport/FAQs#pc_cve_recordscve_record_descriptions_created

of the words and tokens contained in the dataset, we created word clouds for the dataset, which can be seen in Fig. 8.1. The first image represents individual words, the second highlights bigrams, and the third illustrates the tokenized dataset. Across all word clouds, a prominent association with the security domain is evident.

Table 8.2 Statistics of the number of tokens and entries based on the training dataset

#Tokens	Min[a]	Max[a]	Sum	Median	Mean	Entries
Blogs	44	0.1M	169M	710	1.1k	151k
arXiv	533	0.7M	167M	8.2k	9.9k	16k
NVD	5	1.9k	12M	58	71	171k
Twitter	1	500	179M	39	45	4M
Total	1	0.7M	528M	40	122	4.3M

[a] Minimal or maximal token per entry.

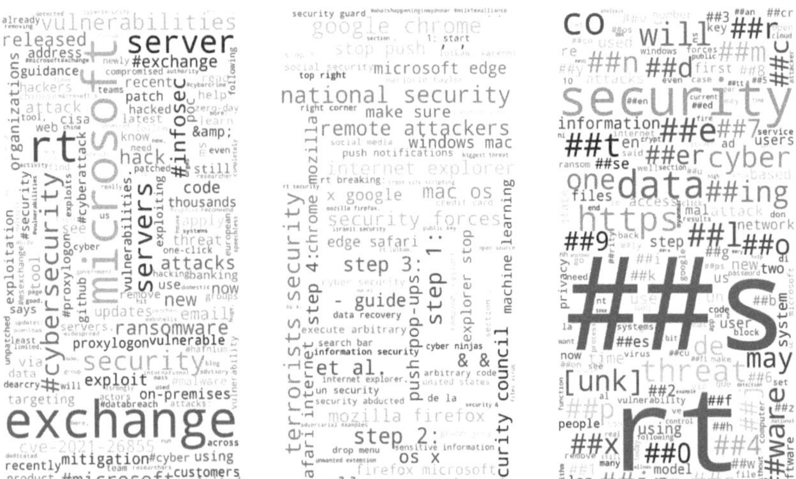

Fig. 8.1 Visualization of the training corpus through word clouds: (a) single words, (b) bigrams, and (c) tokenized input for the model

8.3.3 Architecture

As explained before, we adapt the domain of the BERT model with further pre-training. The architecture is based on the transformer architecture, introduced by Vaswani et al. [46]. Transformers leverage the self-attention mechanism which enables the model to weigh the significance of each word in a sequence relative to the others, thereby capturing long-range dependencies and contextual information. They consist of encoders and decoders, but BERT only employs the encoder portion, *i.e.*, it only encodes the text into an internal representation. During pre-training, BERT utilizes masked language modelling by randomly masking a certain percentage of words in the input sequence. The model is then tasked with predicting the masked words based on their surrounding context. This encourages BERT to develop a deep understanding of the language structure, as it has to infer the masked words using the available context.

From this follows the mathematical representation of the key components: Given an input text which is transformed into a sequence of tokens $X = \{x_1, x_2, ..., x_n\}$. From them, the token embeddings, *i.e.*, fixed-size vector representations (with positional encodings) $E = \{e_1, e_2, ..., e_n\}$ are obtained. For each token, we compute the Query (Q), Key (K), and Value (V) vectors by projecting the embeddings using learned weight matrices WQ, WK, and WV:

$$Q_i = e_i * WQK_i = e_i * WKV_i = e_i * WV \tag{8.1}$$

Next, we calculate the attention scores for each token pair (i, j) (with d being the embedding dimension):

$$\text{score}_{i,j} = \frac{Q_i * K_j^T}{\sqrt{d}} \tag{8.2}$$

Then, we apply the softmax function to the scores to obtain attention weights:

$$\text{weights}_{i,j} = e^{\text{score}_{i,j}} / \sum_{k=1}^{n} e^{\text{score}_{i,k}} \tag{8.3}$$

Finally, we compute the self-attention output for each token by taking a weighted sum of the Value vectors:

$$\text{output}_i = \sum_{j=1}^{n} (\text{weights}_{i,j} * V_j) \tag{8.4}$$

BERT uses multiple self-attention heads to capture different aspects of the context. Each head computes its own self-attention output $output_{h,i}$, which are concatenated and projected using another learned weight matrix WO:

$$\text{multiheadoutput}_i = [\text{output}_{1,i} \oplus \text{output}_{2,i} \oplus \ldots \oplus \text{output}_{H,i}] * WO \tag{8.5}$$

Then, BERT applies a position-wise fully connected feed-forward network to the multi-head attention output, with W1, b1, W2, and b2 as learned weight matrices:

$$\text{nnoutput}_i = \text{ReLU}(\text{multiheadoutput}_i * W1 + b1) * W2 + b2 \tag{8.6}$$

Each encoder layer in BERT consists of the multi-head self-attention mechanism, followed by layer normalization, the position-wise feed-forward network, and another layer normalization:

$$\text{outputmha}_i = \text{LayerNorm}(\text{multiheadoutput}_i + e_i) \tag{8.7}$$

$$\text{outputnn}_i = \text{LayerNorm}(\text{nnoutput}_i + \text{outputmha}_i) \tag{8.8}$$

This process is repeated for each encoder layer in the model, as depicted in Fig. 8.2.

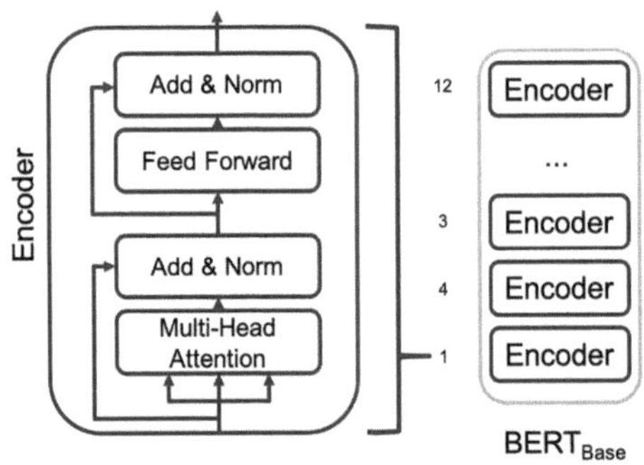

Fig. 8.2 Visualization of the BERT encoder stack [18]

While these mathematical representations describe the core components of the BERT architecture, including self-attention, multi-head attention, and the position-wise feed-forward network within the encoder layers, further details can be extracted from the work by Devlin et al. [47].

8.3.4 Preliminary Evaluation: Catastrophic Forgetting

Catastrophic forgetting describes the phenomenon that an already trained model tends to catastrophically forget the previously trained knowledge when trained on further data [205]. If catastrophic forgetting occurs, our model might lose its ability to understand or interpret nuances and contexts that may not be explicitly covered in the cybersecurity domain but are still relevant for processing cybersecurity-related texts. (1) In real-world applications, data often contains a mix of domain-specific and general language. For instance, cybersecurity texts may include general language expressions, analogies, or references that require a broader understanding of language beyond just technical terms. (2) There are certain tasks where the model has to handle in domain and out of the domain instances. Like, for example, when the task is in a intersection of cybersecurity and another domain (like law, policy, and management). A model exclusively trained on cybersecurity might struggle with texts that blend these disciplines. A model that retains its general language capabilities alongside its domain-specific knowledge is more robust and flexible [206]. Finally, the work of et al. [207] shows that avoiding catastrophic forgetting in domain-specific BERT models even improves the performance on the domain tasks. This illustrates the benefits of a well-balanced model that can handle both domain-specific and general language data effectively.

Research [47, 208] suggests that the second training should be more lightweight than the first, so that the already trained knowledge is not overshadowed. In this regard, the three most important hyperparameters are the learning rate, which determines how much the new data updates the weights of the model, the epochs, *i.e.*, how long the model is trained and how many updates are made, and finally the size of the dataset. The work of Sun et al. [208], for example, addresses the problem of catastrophic forgetting by examining different learning rates, demonstrating that a lower learning rate is best. Besides, the authors of the BERT paper Devlin et al. [47] recommend training with only a few epochs. Accordingly, further training can often lead to overfitting of the data and, as already discussed, to catastrophic forgetting. This remark can also be transferred to the size of the datasets for DAPT, as they are very large and correspondingly many training updates are carried out. The difficulty is to find the right configuration of these parameteres to avoid catastrophic forgetting and at the same time to enable enough learning that the model achieves high quality in the new domain. Therefore, we test different configurations of the hyperparameters' learning rate, dataset size, and number of epochs in this preliminary evaluation. Since training a model already requires a considerable amount of time and resources (about 4 days on 4 Nvidia V100 GPUs), we cannot perform an extensive hyperparameter search, such as the bio-inspired search by Ibor et al. [209]. Hence, we narrow down the range of configurations by taking into account the research findings and by orienting our approach towards the original BERT training process. The authors of BERT trained the model with a learning rate of 1×10^{-4} for about 40 epochs. We therefore propose the following configurations:

Training Configuration

- Learning Rate: 1×10^{-4}, 2×10^{-5}, 5×10^{-5}
- Epochs: 20, 30, 40
- Data: 5%, 10%, 15%

For the other hyperparameters, we followed the original BERT approach, *i.e.*, we used a weight decay of 0.01, a dropout rate of 0.1, 10000 warm-up steps, and ADAM as the optimisation algorithm [210]. We trained all models with a batch size of 64 on 4 Nvidia Tesla V100 GPUs.

To be able to measure the degree of forgetting, as well as the performance in the security domain, we evaluate the different models on the SuperGLUE task BoolQ and the security task MSExchange (a more detailed description of the datasets follows in Sect. 8.4.1).

The results are presented in Table 8.3. As can be seen, almost every configuration performs well in both tasks. The learning rate of 2×10^{-5} appears to be better in both tasks, which is in line with the results of Sun et al. [208]. Similarly, 30 epochs perform better on both tasks than the higher and lower configurations. However, with regard to the size of the dataset, the results are not so clear. While a dataset size of 10% of the BERT training dataset size performs best on the MSExchange dataset, a dataset size of 5% of the original BERT training dataset size seems to perform best on the BoolQ task. This is not surprising, as the degree of forgetting is lower with less new training data. Since the model with 10% still performs very well on the BoolQ task and because we weight the performance in the security domain more heavily, we decided to use the model with 10% of the original BERT training dataset. The training loss of this CySecBERT model can be seen in Fig. 8.3. Accordingly, the loss decreases logarithmically and improves only very slowly after 300k steps.

Table 8.3 Pre-evaluation of different CySecBERT configurations considering the catastrophic forgetting task and the performance in the security domain with the boolq task (accuracy) and the MSExchange task (F1), respectively

Configuration	BoolQ	MSExchange
LR 2×10^{-5}, Epochs 30, Data 10%	0.6752	**0.8869**
LR 1×10^{-4}, Epochs 30, Data 10%	0.6722	0.8620
LR 5×10^{-5}, Epochs 30, Data 10%	0.6700	0.8544
LR 2×10^{-5}, **Epochs 20**, Data 10%	0.6660	0.8860
LR 2×10^{-5}, **Epochs 40**, Data 10%	0.6614	0.8846
LR 2×10^{-5}, Epochs 30, **Data 5%**	**0.6785**	0.8816
LR 2×10^{-5}, Epochs 30, **Data 15%**	0.6716	0.8496

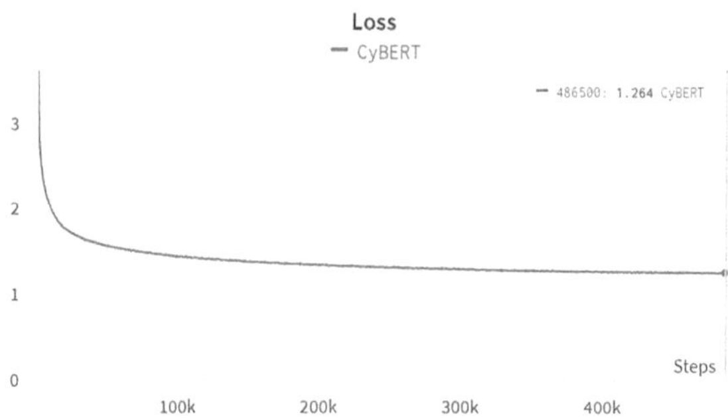

Fig. 8.3 Training loss of the CySecBERT model

8.4 Evaluation

In this section, the evaluation process and the corresponding results are presented in detail. We cover a short description of the evaluation tasks in Sect. 8.4.1, followed by the presentation and interpretation of the results in Sect. 8.4.2.

8.4.1 Experiments and Tasks

Since our goal is to publish a model that is highly usable for the cybersecurity domain, we evaluate it against the current standard method of the domain (BERT)

and against another cybersecurity language model (CyBERT from Ranade et al. [36]). We use different types of tasks, *i.e.* intrinsic and extrinsic evaluation tasks, which are reasonably chosen for the field of cybersecurity. While the extrinsic tasks measure how well the trained model performs on downstream tasks, *i.e.* measure real-world application, the intrinsic tasks measure the model itself without any kind of additional classifier, *e.g.* by measuring the representations of the model and by demonstrating an overall fit to the domain.

As intrinsic tasks, we apply a word similarity task using parts of the dataset by Mumtaz et al. [195] and a Twitter dataset for clustering evaluation. The clustering dataset is based on a random sample of Log4j Twitter posts. For this task, posts are converted into latent representations with different BERT models. The latent representation consists of the concatenation of the last four layers of the model output and of the mean values across all words in the post. A K-Means clustering algorithm with k-values from 5 to 9 is executed on the gathered and transformed posts. The evaluation scores are measured with the Silhouette Coefficient (the higher, the better). This is an internal clustering metric that analyzes the clusters created and does not require gold labels. It is important because there can be many solutions and gold labels can be misleading in the case of clustering [211]. The cybersecurity word similarity dataset consists of words with their equivalents, all from the field of cybersecurity. Moreover, the dataset is an extension of the public cybersecurity word similarity dataset from Mumtaz et al. [195] and contains over 300 word pairs. For this extension of the original dataset, a cybersecurity expert followed the process of Mumtaz et al. [195] and added more cybersecurity words that can be considered similar. Normally, the word similarity evaluation is based on the cosine similarity of the word embeddings when static embeddings are used. The problem is that BERT is context-dependent, which means that a clear word embedding cannot be given without context and the standard method of measuring word similarity is no longer suitable. For this reason, we have developed a new method for evaluating word similarity, in which the model predicts whether two given words are similar. Similar to the works on zero-shot learning, we create a meaningful cloze task [212] consisting of a sentence with a masked word that the model fills in, which is implicitly the answer to the similarity question. The task is written in the following way:

Explanation Cloze Task [212]
"*Are* $word_1$ *and* $word_2$ *similar? [MASK]*", *where* **[MASK]** can be either "Yes" or "No", which represent the masked words that the model has to fill.
Example: "*Are* ***virus*** and ***malware*** similar? **[MASK]**"

This trick of zero-shot learning means that no explicit classification model needs to be trained, and the task remains intrinsic. Besides, we aim to show that the evaluated model does not only predict the similarity of each word to all other words, but also detects when two words are not similar. We therefore randomly take word *Paris* from the dataset that are not similar and add them to the evaluation. This dataset is then used for evaluation and an F1 score is calculated for the model's predictions.

As part of the extrinsic tasks, we employ two cybersecurity classification tasks from Riebe et al. [11] and Bayer et al. [32]. In the first task, the classifier has to decide whether a Twitter post is related to the field of cybersecurity, and in the second task, it has to predict whether a post might be relevant to experts in the field during a major cybersecurity event. Furthermore, we use the sequence tagging dataset by Kuehn et al. [5]. Sequence tagging is the task of finding specific words in a text, and it requires that each word in a text is tagged, often as IOB: Either *I-nside*, *o-utside*, or *b-eginning*, referring to the specific words being searched for. Specifically, the dataset consists of several (NER) tasks (recognition of named entities) for predicting relevant details of NVD descriptions. We then chose the task of predicting the name and version of the software and the attack vector. We decided for the employment of these tasks because of their different performances in Kuehn et al. [5]'s work, in order to analyze how well the models perform on varying levels of difficulties of the tasks.

Furthermore, we evaluate CYSECBERT and BERT on the SuperGLUE benchmark [213]. This is a common NLP benchmark which we utilize to identify signs of catastrophic forgetting [106]. We assume that our cybersecurity model is not able to achieve a better or even equivalent result, as a certain degree of forgetting is acceptable and necessary for learning. Nevertheless, we expect the performance not to be too poor, as this would otherwise indicate that some basic knowledge would have been forgotten during the domain training phase.

Neural networks consist of many random hyperparameters that have to be fixed before each training and that are not transferable to other learning processes. As shown by Reimers and Gurevych [214], the frequent execution of a training process can be used to avoid incorrect inferences arising from a randomly better choice of hyperparameters. Accordingly, and following the experiments of Sanh et al. [215] and Liu et al. [198] all extrinsic experiments were performed five times and the mean values as well as the standard deviation are given in the respective tables.

8.4.2 Results

As stated in Sect. 8.4.1, we aim to evaluate CySecBERT, as well as BERT and CyBERT [36] on different tasks which are mainly located in the cybersecurity domain, incorporating both intrinsic and extrinsic evaluation tasks. Additionally, we run the SuperGLUE task to test our model for catastrophic forgetting.

8.4.2.1 Intrinsic Tasks

To measure the representation quality of the model, we evaluate it by applying two intrinsic tasks from cybersecurity: clustering and word similarity.

Clustering The results of the clustering task are given in Table 8.4. While the CyBERT model of Ranade et al. [36] only performs better than the baseline when forming 5 to 7 clusters, our approach performs better in every constellation, according to the Silhouette Score. In fact, CySecBERT outperforms Ranade et al. [36]'s cybersecurity model by a considerable margin for each number of clusters, with consistent improvements ranging from +0.002 to +0.059 points. Our model shows the highest improvement when 9 clusters are formed. On the one hand, these results demonstrate that we can obtain more coherent clusters thanks to our trained language model. On the other hand, from a more general perspective, the results show that the model is more able to represent the given instances in a meaningful latent space.

Table 8.4 Silhouette Score of the first intrinsic task, clustering the data of a Log4J dataset. The best values are marked

# Clusters	BERT	CyBERT [36]	CySecBERT
5	0.114	0.141	**0.143**
6	0.115	0.124	**0.150**
7	0.118	0.133	**0.167**
8	0.125	0.117	**0.163**
9	0.130	0.113	**0.172**

Nevertheless, even better results could be expected if we used an approach like SentenceBERT by Reimers and Gurevych [45] for our model, as they have proven to be much more suitable for representing complete documents, like tweets in our case.

Word Similarity The word similarity task results are displayed in Table 8.5. The baseline model has the worst performance with a F1-score of 0.44. This is to be expected, as most domain-specific words were not or only very rarely included in the standard BERT training. Our CySecBERT model is clearly superior to the other two approaches, which is remarkable as it confirms the previous intrinsic results. However, we would like to point out that this task is different from other word similarity tasks as it does not reflect word similarities directly through the word representations, but by questioning the model in a cloze fashion Sect. 8.4.1.

Table 8.5 Overview of the results of the word similarity task where the scores are indicated by the F1 score

Tasks	Word Similarity
BERT	0.4382
CyBERT [36]	0.4861
CySecBERT	**0.6382**

8.4.2.2 Extrinsic Tasks

After we have already demonstrated that the model produces meaningful representations of cybersecurity-related vocabulary and data, we want to test whether our model is also comparably more suitable for real-world applications, *i.e.*, for extrinsic tasks, the so-called downstream tasks of machine learning. The tasks that we have chosen for the cybersecurity domain are (i) Named-Entity Recognition (NER) (ii) general relevance classification, and (iii) CTI classification.

NER The results of the NER task are shown in Table 8.6. The table shows that the results of the original work of Kuehn et al. [5] are worse than those of the methods evaluated in our study. While the basic BERT model and the CyBERT model of Ranade et al. [36] are more similar to each other, *e.g.*, in terms of software naming (SN), our model consistently outperforms both. Only in the tagging of the software version (SV) does the CyBERT model of Ranade et al. [36] perform significantly better than the baseline BERT model, whereas our model nevertheless improves this result. Hence, it could be speculated that the CyBERT training data from Ranade et al. [36] contained entries of software versions at a higher frequency than the normal BERT data. However, the CyBERT model deteriorates the results for software names (SNs), which could indicate that either numerous software names are missing in high frequency in their dataset, or that they have been neglected due to errors in the training process. The highest improvements of our model can be seen in attack complexity (AC), outperforming the CyBERT model of Ranade et al. [36] by with 0.0136 points. Nevertheless, we perceive the results of this particular task as overall not very satisfactory. Reasons therefore have already been discussed by Kuehn et al. [5] and are mainly related to the problem of too little data in this task.

Table 8.6 NER shore based on tagged software versions (SVs), software names (SNs), and attack complexity (AC) of NVD descriptions. The results are given as F1 scores and the best values are marked

CVSS NER	SV	SN	AC
[5][a]	0.8735 (–)	0.8584 (–)	– (–)[b]
BERT	0.9247 (0.0064)	0.8837 (0.0037)	0.3323 (0.0135)
CyBERT [36]	0.9298 (0.0019)	0.8834 (0.0029)	0.3336 (0.0214)
CYSECBERT	**0.9302** (0.0066)	**0.8871** (0.0025)	**0.3472** (0.0116)

[a]Showing the reported results of the work.
[b]No comparable results available.

Relevance Classification (CySecAlert) In the first classification task of our experiments, the models are trained to predict whether a Twitter post is related to the cybersecurity domain (*cf.* Table 8.7). This can be considered a general cybersecurity task, as the model only has to identify cybersecurity-related words. The original work by Riebe et al. [11] has the worst results, possibly due to the use of a classical machine learning method. It is particularly interesting to see that the Ranade et al. [36] model performs worse than the basic BERT model. Our model significantly improves the base model and the CyBERT model by 0.0104 and 0.0236 points in the F1 score, respectively. All models have a relatively low standard deviation, indicating that the fine-tuning process is stable across all runs.

Specialized Cyber Threat Intelligence (CTI) Classification (MS Exchange) The second classification task is about identifying specialized CTI where very specific words are needed to classify the instances. The results of this task are also presented in Table 8.7. Surprisingly, unlike in the previous tasks, the CyBERT model of Ranade et al. [36] is able to improve the baseline, showing that while it does not contribute improvements in the more general tasks, it could be beneficial in more specific tasks. There is a high improvement of our model compared to the baseline observation (+0.027), which we expected since this task focuses on very domain-dependent language and specific vocabulary. Moreover, although the CyBERT model of Ranade et al. [36] is advantageous for this task, our model still improves the results significantly by +0.0103 F1 points. Besides, it is observable that our model has a significantly lower standard deviation than the other two models, which again indicates a very stable training process. Table 8.7 also contains the result of the work by Bayer et al. [32], which is very similar compared to the BERT baseline, but the best standard deviation.

Table 8.7 Classification results of the MS Exchange and CySecAlert dataset, given as F1 scores. The best values are marked

	MS Exchange	CySecAlert
[11][a]	–	0.8051 (–)
[32][b]	0.8536 (0.0007)	–
BERT	0.8599 (0.0193)	0.8779 (0.0084)
CyBERT [36]	0.8766 (0.0153)	0.8647 (0.0095)
CySecBERT	**0.8869** (0.0026)	**0.8883** (0.0064)

[a] Showing the reported results of the work.
[b] Showing the reported results of the work.

Furthermore, we also included the few-shot learning task of this dataset in Table 8.8, where only 32 instances are given as training data. CySecBERT is an integral part of the few-shot approach of Bayer et al. [32]. The results show a significant improvement over the baseline and over the state-of-the-art few-shot learning method ADAPET by Tam et al. [216]. This demonstrates how the CySecBERT model is incorporated and applied in further research, contributing to the advancements in other disciplines. Details of the approach and further results can be extracted from the work of Bayer et al. [32].

Table 8.8 Classification results of the few shot experiments of the MS Exchange dataset. The best values are marked

Name	Model	Accuracy	F1
Baseline	BERT	0.4965 (0.0190)	0.5870 (0.1881)
ADAPET	BERT	0.6589 (0.0135)	0.6254 (0.0432)
[32]	CySecBERT	**0.7913** (0.0056)	**0.8063** (0.0027)

8.4.2.3 Catastrophic Forgetting

In the final part of our evaluation, we revisit the problem of catastrophic forgetting. To this end, we evaluate our model with the SuperGLUE benchmark to test if the model degrades the results considerably, which would indicate that the model has forgotten the initial knowledge acquired in the BERT training phase. The results of this task and a comparison to the BERT model is displayed in Table 8.9. As expected, we can see that our model reduces almost every task outcome. Nevertheless, the worsening of results are not an indication for catastrophic forgetting, as the differences are still relatively small, with a mean drop of about −0.05 points. This shows that although the model has lost some of its knowledge, most of it is still present. Besides, it is particularly interesting that the performance in the cb task has even increased and the result of the boolq task has remained almost the same.

Table 8.9 Results of the SuperGLUE benchmark, each indicated in the evaluation metric proposed in the benchmark

	Record	RTE	WIC	WSC	Boolq	CB	Copa	Multirc	Total
BERT	0.6416	0.5949	0.6476	0.5538	0.6760	0.3704	0.606	0.4067	0.6010
CyBERT [36]	0.6346	0.6173	0.5980	0.5337	0.6887	0.5676	0.615	0.4146	0.6065
CYSECBERT	0.6137	0.5545	0.5887	0.5404	0.6752	0.5551	0.486	0.3915	0.5468

In addition, we have again included the results of the CyBERT model by Ranade et al. [36], which, interestingly, performs almost the same as the BERT model. One could interpret this as an indication that the model did not suffer catastrophic forgetting during training. However, taking the results of the extrinsic and intrinsic domain tasks into account, we can conclude that their model might have learned very little overall, as it is seldom better, sometimes worse and most often on par with BERT. This can be attributed to the training process of Ranade et al. [36], who trained their model only on 17,000 documents with one epoch, whereas our model is trained on 4,300,000 documents with 30 epochs.

8.4.2.4 Conclusion

In the evaluation, we have shown that the model developed in this work is very well adapted to the cybersecurity context. The tasks have demonstrated that the CYSECBERT model is able to outperform the BERT baseline and the CyBERT model by Ranade et al. [36] consistently across all cybersecurity tasks. We evaluated these models on intrinsic cybersecurity tasks where we summarized how accurate the models represent documents and words in latent space. Based on these tasks, the fundamental quality of the language model has been assessed. In addition, we evaluated the three models using extrinsic cybersecurity tasks that demonstrate the practicality of the model in most real-world application contexts. Our model improves the results of these tasks by up to 0.027 F1 points compared to the other two models, and it achieves its highest improvement on an in-depth cybersecurity task where very specific language differences have to be considered. Moreover, we analyzed the phenomenon of catastrophic forgetting by evaluating our model on standard NLP tasks. Although we observed a deterioration in performance in these tasks, it is only within the expected range of decline. Besides, we concluded that the CyBERT model of Ranade et al. [36] may have been trained on a much too small corpus with too few training steps (CyBERT: 17,000 documents and one epoch versus our CySecBERT: 4,300,000 documents and 30 epochs). We can say with confidence that our model is capable of handling a wide range of cybersecurity tasks while retaining the original language modelling knowledge.

8.5 Discussion, Conclusion, and Outlook

In this work we propose a novel state-of-the-art cybersecurity language model based on BERT [47]. We performed DAPT on this model with a sensibly chosen cybersecurity corpus. The corpus consists of a variety of source data structures, such as blogs, scientific paper, as well as Twitter data. The data and the sources were selected to be appropriate for cybersecurity research and practice. Furthermore, the size of the dataset and the structure of the training process were chosen to prevent catastrophic forgetting on the one hand, and, on the other hand, to enable enough learning for the model to contribute to the general field and to the specific niches of cybersecurity. We explored this through a pre-evaluation phase by tuning the hyperparameters in terms of catastrophic forgetting and domain quality. We then highlighted the performance of our model with a thorough main evaluation of various tasks and drew a comparison to the BERT baseline as well as the current state-of-the-art of cybersecurity language models. First, we evaluated the models on two intrinsic tasks, where we demonstrate that the quality of our model improves in terms of the learned representation space, *i.e.*, how well the cybersecurity-specific instances (words and texts) can be distinguished from each other. Table 8.4 and 8.5 show the substantial performance increases by our model. Second, we evaluated the model together with the other two models for cybersecurity-specific classification and NER tasks to exemplify the usefulness and practicality of the model in application contexts. Our model outperforms the other models in every task, which can be derived from Table 8.6 and 8.7. The greatest improvement is observed in the special CTI classification dataset, suggesting that the model is particularly beneficial when dealing with very specific cybersecurity language that is not contained in the training dataset of the standard BERT model. Our evaluation concludes with a focus on catastrophic forgetting by which we assessed the performance of our model against a general NLP benchmark. While these results (Table 8.9) point out that our model does indeed degrade them overall, they also show that, as intended, there is no catastrophic forgetting and that the final model has combined much of its original knowledge with the new knowledge about cybersecurity.

While we are aware that the current state-of-the-art on research in language modeling and NLP generally tends to focus on larger language models like GPT-3 by Brown et al. [52], we have chosen the BERT model on purpose. Most of the cybersecurity research and especially practice does not have the necessary resources to apply large language models. In most cases, the BERT model can still be considered the standard model in ML application contexts such as the cybersecurity domain. In this way, our work is most beneficial to the research landscape and to practice.

Practical and Theoretical Implications Our work contributes to research and pratice through a novel, state-of-the-art cybersecurity model called CYSECBERT, which we have published. We also publish the associated dataset so that it can contribute to further research. Thus, our work has several implications for practice and research:

A novel, state-of-the-art language model for cybersecurity that is useful for various tasks. With our research surrounding the model, we have aimed to find a solution to increase the performance of machine learning in cybersecurity language tasks. Our model provides utility for numerous tasks, as can be estimated based on the success in extrinsic task scores as well as inferred from intrinsic task scores. Thereby, they show that the representation space is better for the domain-dependent language with our model. With the release of our model, we are paving the way for better cybersecurity tools, as practitioners can easily use the new model in existing pipelines, for example, in alert aggregation [73], in detection of phishing websites [217, 218], or even in malware detection Bayer et al. [219]. More advanced tools will then also be the result of new research derived from the model and have the potential to improve the results in various tasks by incorporating further research ideas, as has already been implemented as part of the research of Bayer et al. [32], for example, where the CYSECBERT model has been integrated. This can be pursued on a smaller scale, where the model is not the focus but serves as a foundation on which further techniques such as data augmentation, meaningful data selection, few-shot learning or specific applications are built. Yet, it can also be done on a larger scale, where the model is the subject of research, for example by analyzing its results in explainable AI approaches.

In our evaluation, we demonstrate that the CySecBERT model can serve as a substitute for other models in the field of cybersecurity by providing new state-of-the-art performances. However, the CySecBERT model will not be suitable as a replacement for every type of cybersecurity model. We expect that further work in this area will adopt our methodology and will train other models that may be even more specialized regarding a specific cybersecurity topic or that may have a much larger neuron size.

Evaluation of catastrophic forgetting in terms of learning rate, dataset size and number of epochs. When training the CySecBERT model, we paid special attention to catastrophic forgetting phenomena where the trained model loses its valuable initial knowledge. To ensure that the level of catastrophic forgetting is kept to a minimum, we first performed a preliminary evaluation of the hyperparameters' learning rate, dataset size and number of epochs, as research in this area suggests that the second training process should not overshadow the first training [47, 208]. We narrowed down the space of possible hyperparameter constellations by these considerations and evaluated seven fully trained CySecBERT models on a standard NLP and on a cybersecurity task to measure the degree of catastrophic forgetting and the quality in the cybersecurity domain. Our results are in line with research and show that the training process should not be too heavyweight, *e.g.*, by heavy updates due to a high learning rate, but also not too lightweight, so that domain knowledge is not acquired, *e.g.*, due to a smaller dataset. In the final section of our main evaluations, we evaluated the CySecBERT model on the SuperGLUE benchmark, which consists of standard NLP tasks, and compared it with the BERT model and the CyBERT model of Ranade et al. [36]. While the results show that our model performs worse than the BERT model, this is to be expected since some level of catastrophic forgetting will always occur when further training the model towards a certain domain. However, it is interesting to note that the results of the CyBERT model of Ranade et al. [36] are very similar to the BERT model, which is due to the small training process which the CyBERT model originates from. While the CyBERT model is trained on only 17,000 documents with one epoch, our model is trained on about 4,300,000 documents with 30 epochs.

It is not clear whether these results, especially with regard to the hyperparameters, are generalizable to other language models, especially to very large language models such as GPT-3 and 4. For an evaluation with a broader hyperparameter search, as for example executed with a bio-inspired, evolutionary method from Ibor et al. [209], or for a focus on these very large language models, we expect research to be primarily concerned with catastrophic forgetting and with having the necessary extensive resources.

A cybersecurity dataset containing most relevant sources for the training process. The dataset was created with consideration given to including a variety of sources and textual types. This ensures that the model can be applied to a broad range of cybersecurity tasks. However, we anticipate future work analyzing the published dataset, for example, to enhance its quality based on specific criteria or to identify any unintended biases. The dataset can also serve as a basis for training other language models. Although we have deliberately chosen this size of the dataset for BERT training to prevent catastrophic forgetting, it might be useful to expand it, which can easily be done by collecting more data from the sources that we have already selected. It might also make sense to utilize larger language models that might achieve even greater performance or that might even use neural architecture search to find a suitable architecture for special use cases, as proposed by Shang et al. [220] or Okunoye and Ibor [221].

Ethical Considerations We would like to emphasize that we did not explicitly focus on and analyze social biases in the data or the resulting model. While this may not be so damaging for most application contexts, there are certainly applications that depend heavily on these biases, and including any kind of discrimination can have serious consequences. As authors, we would like to express our warnings regarding the use of the model in such contexts. Nonetheless, we aim for an open source mentality, observing the great impact it can have, and therefore transfer the thinking to the user of the model, drawing on the many previous discussions in the open source community.

Bandit on the Hunt: Dynamic Crawling for Cyber Threat Intelligence

9

Abstract

Public information contains valuable Cyber Threat Intelligence (CTI) that is used to prevent attacks in the future. Ideally, the learnings from previous attacks help to mitigate all those that follow. While there are standards for sharing this information, much of it is shared in non-standardized news articles or blog posts. It is a time-consuming task to monitor online sources for threats and even then, one can never be sure, to use the right sources. Current research propose extractors of Indicator of Compromise (IoC) from known sources, while the identification of new sources is rarely considered. This paper proposes a focused crawler focused on the CTI domain based on multi-armed bandit (MAB) and different crawling strategies. It uses Sentence-BERT (SBERT) to identify relevant documents, while dynamically adapt its crawling path. We propose a system called THREATCRAWL, which achieve a harvest rate of over 25% and is able to expand its used seed by over 300%, while retaining focus on the topic at hand. In addition, this crawler identified previously unknown but highly relevant overview pages, datasets, and domains.

Original Publication ***Kuehn, P.***, *Nadermahmoodi, D., Bayer, M., Reuter, C., "Bandit on the Hunt: Dynamic Crawling for Cyber Threat Intelligence,"* arXiv, *Apr. 25, 2025.* https://doi.org/10.48550/arXiv.2504.18375.

Minor changes regarding language, style, and formatting have been made in this chapter to provide a consistent experience.

P. Kühn, *Proactive Cyber Threat Intelligence*, Technology, Peace and Security | Technologie, Frieden und Sicherheit,
https://doi.org/10.1007/978-3-658-49241-0_9

9.1 Introduction

In CTI information is used to learn from current threats and prevent attackers from using similar methods against infrastructures. This is done by sharing actionable information such as IoC through various channels. The CTI already established standards to publish, import and export this information to CTI databases and applications. But CTI is often shared in an unstructured way, such as blog posts or threat reports from security companies or experts [57, 222]. For personnel, it is a time-consuming and mundane task to manually scan online posts for IoC. Hinchy [27] surveyed 468 full-time security analysts about their work. Over the half of participants (i) spend over half their time on manual tasks, (ii) believe over half of their tasks can be automated, and (iii) are likely to change jobs in the next year if modern and automated tools to support them are not being developed. Kaufhold et al. [28] confirmed these results with participants stating that they "do not have the capacities to monitor all media" and that they "would benefit from a higher degree of automation".

IoC extractors and threat detection methods are already being developed and tested [33, 223], as well as vulnerability severity predictors [6, 224]. However, this field of research assume perfect information sources. But the Internet is a dynamic environment, where sites change focus, stop publishing, or new sites are added. In the current situation, manual identification of new sources and the already present information overload at security personnel contradicts each other.

It requires manual and, above all, unavailable work to maintain an up-to-date set of CTI-relevant websites. This leads to blind spots of currently active attack campaigns. In addition, the amount of time spent searching for CTI leads to less effective and motivated staff [28]. Ultimately, this problem reduces infrastructure security because less time is spent actually securing the infrastructure. To avoid this, it is necessary to develop a new and precise approach to finding relevant web pages quickly and effectively. Focused crawlers "selectively seek out pages that are relevant to a pre-defined set of topics" [225]. This work focuses on the identification of CTI-related information published in the web. We aim to answer the following research question *"How can CTI related information be identified and crawled from the web?"* (**RQ**)

This work combines different techniques for crawling, classifying, ranking content and proposes a focused crawling pipeline called ThreatCrawl. It allows gathering specific information of the CTI domain from the surface web. Our proposal uses-BERT (SBERT) embeddings for websites to decide whether to follow sources (**C1**). Such documents are classified by type of information and ranked according to their suitability in the domain (**C2**).

First, we provide an overview of related work in Sect. 9.2. Section 9.3 presents the theoretical concept of ThreatCrawl, followed by a description of the initialization in Sect. 9.4, while Sect. 9.5 evaluates the system. Section 9.6 discusses the evaluation, names limitations and future work, and Sect. 9.7 concludes this work.

9.2 Related Work

Traditional document embeddings, such as Term Frequency-Inverse Document Frequency (TF-IDF), are long surpassed by context aware, deep-learning based embeddings [47, 226]. [45] build on Bidirectional Encoder Representations from Transformers (BERT) by proposing a method for generating sentence embeddings, called SBERT, which is also suitable for documents.

Several focused crawlers have been proposed using TF-IDF as embedding method [227–229]. Zhang et al. [40] proposes a method to find datasets lacking metadata, and hence, are not included it Google Dataset Search. This builds on a MAB focused crawler using keywords for focus [229]. Koloveas et al. [41] proposes an integrated machine learning (ML)-based crawler framework for managing and sharing CTI information, using the ACHE for crawling and Gensim for relevance classification. Sanagavarapu et al. [230] proposes a cybersecurity specific search engine.

While a lot of CTI research suggests focusing on Twitter/X, because it provides easy access to cyber threat information [11, 231], others use a set of known sites [33]. While both methods produce valid studies, they are based on known sources, either in the form of Twitter/X handles or websites. There is no way to expand the view. And since the leadership change of Twitter/X, crawling it became more difficult. Dekel et al. [232] use a MAB approach to prioritize investigated attacks in a CTI dataset.

Current work that combines web crawling, classification, and page ranking techniques for CTI in a single pipeline is vastly discussed. [50] describe a crawler, which uses SBERT embeddings to label downloaded documents. With this approach, crawling and classifying are two completely independent processes, *i.e.*, everything is crawled and then classified. Koloveas et al. [41] present a different two-step approach to the crawling and classification of their documents, reducing the benefits of focused crawling (saving hardware and network resources). Their technique show a *harvest rate* of 9.5%, *i.e.*, 9.5% of crawled pages are of actual interest. Computer Emergency Response Team (CERT) and Security Operations Center (SOC) personnel do already have a small set of regularly monitored domains, which inherently defines the scope of their infrastructure. Thus, an open question is the development of a general one-step focused crawling approach that combines the crawling, classification, and ranking of CTI-relevant content, based on a known set of URLs.

9.3 Methodology

The goal of the present work is the identification of new, previously unknown web pages, that are related to the user's input URLs. We combine the different concepts proposed in [229], integrate new technology, and tailor them to the requirement of security personnel.

9.3.1 Notation

We denote $\mathbb{P}$ as the set of all web pages. Given two pages $p, p' \in \mathbb{P}$, $p \rightarrow p'$ indicate, that p links to p' and $p \sim_\theta p'$ indicates that p is contextually similar to p' with regard to a relevance threshold θ. $theta$ is omitted if it is clear from context. We extend $\sim$ to sets, *i.e.*, $p \sim_\theta P \subseteq \mathbb{P} \iff \exists p' \in P.p \sim_\theta p'$. Similarly, $P, P' \subseteq \mathbb{P}$, $P \sim \theta P' \iff \forall p \in P.p \sim \theta P'$. The function $crawl_a(p) \rightarrow \mathcal{P}(\mathbb{P})$ denotes a crawl of page $p \in \mathbb{P}$ based on a crawl action a, which returns a set of pages.

9.3.2 Problem Definition

One of the key aspects of this system is the relevance of information. Recent studies show a shift of relevance in the CTI domain[1] from very detailed information like IoC to broader information like threat or malware reports (cf. Table 9.1). Therefore, we calculate the relevance of pages based on their similarity to the used seed $s \in S \subseteq \mathbb{P}$, rather than using binary information like the existence of IoC information [33]. Our approach might miss dataset pages that present IoC information, *e.g.*, through simple lists, but this is already covered by [229]. Def. 9.3.1 defines the problem, we aim to solve.

Definition 9.3.1 (CTI-focused crawler) Given a set of seed pages $S \subseteq \mathbb{P}$ defining the scope of ones CTI infrastructure security information, we want to identify $P \subseteq \mathbb{P}$, such that $P \sim S$.

[1] Based on an annual survey conducted by the SANS Institute (https://www.sans.org) in which they survey security professionals from various organizations.

Table 9.1 Top 3 most useful CTI types according to the SANS CTI surveys from the years 2020 to 2022 [233–235]

	2020	2021	2022
#1	IoC	Textual information about targeted vulnerabilities	Textual information of used malware
#2	TTPs	Textual information of used malware	Textual information about targeted vulnerabilities
#3	Adversary analysis	IoC	Broad information about attacker trends

We propose an architecture we call THREATCRAWL. It uses SBERT embeddings, and a one-step approach, *i.e.*, classification during the crawling process to adjust the crawling direction on the fly. It implements similar retrieval methods as [229] combined with a UCB1-MAB [236]. The retrieval methods are (i) forward (F), (ii) backlink (B), and (iii) keyword search (K), *i.e.*, $crawl_a$ with $a \in [F, B, K]$. Given a page $p \in \mathbb{P}$. *Forward search* crawls all links provided on p, *i.e.*, $crawl_F(p) = \{p' \in \mathbb{P} \mid p \rightarrow p'\}$. *Backlink search* crawls all pages, which link to p, *i.e.*, $crawl_B(p) = \{p' \in \mathbb{P} \mid p' \rightarrow p\}$. *Keyword search* searches for pages containing keywords of p, *i.e.*, $crawl_K(p) = \{p' \in \mathbb{P} \mid p' \text{ contains keyword of } p\}$. Those methods are proven to provide a broad coverage during focus crawling [237–239], while still providing the possibility to dive deep in a source through forward searches.

9.4 ThreatCrawl

Our proposed architecture to answer our research question is presented in (*cf.* Section 9.4.1), followed by insights into the search actions (*cf.* Section 9.4.2), classification method (*cf.* Section 9.4.3), and MAB selection (*cf.* Section 9.4.4).

9.4.1 Overview

We propose THREATCRAWL (cf. Def. 9.3.1) to answer our research question *"How can CTI related information be identified and crawled from the web"*. Its central operation is bound on the user-provided set of seed URLs $S \subseteq \mathbb{P}$ and the used crawling methods F, B, and K. An overview of the whole system is presented in Fig. 9.1. S is crawled as ground truth for the classification and to provide the initial steps for crawling and prepare the MAB. This is done using a priority queue based on the relevance of a given page. For the seed pages $p \in S$ the priority is set to the highest possible value. Following this step, the MAB is initialized with a discovery phase to calibrate the arm selection for the crawl methods (i) forward, (ii) backlink, and (iii) keyword search. All pages that are discovered this way can be directly classified as relevant through their content and if relevant, pushed into the priority queue with their similarity to S directly translated as priority, *i.e.*, more similar pages are picked first. Afterward, the priority queue is processed one step at a time, popping the most relevant page and using the MAB to select the most rewarding arm until one of the stop conditions is reached (cf. Sect. 9.3). This ensures the highest possible outcome measured as *harvest rate*, *i.e.*, $\frac{|R|}{|P|}$, where $R, P \subseteq \mathbb{P}$ is the set of relevant pages and the set of all seen pages by the crawler, respectively.

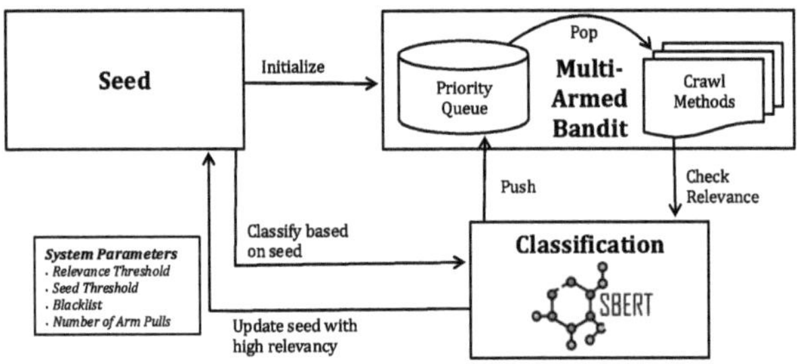

Fig. 9.1 A schematic overview of our proposed CTI source identification system THREATCRAWL leveraging a (MAB) and classification based on Sentence-BERT (SBERT) [45]

9.4.2 Search Actions

Forward Link Search Forward link search follows all hyperlinks on a page p, offering broad website coverage and uncovering in-depth content.

Backward Link Search Backward link search, commonly used in SEO tools, identifies pages linking back to p to discover related content. Since no open sources provide this data, we rely on a commercial API.

Keyword Search Keyword search extracts key terms from p to guide searches, using tools like *KeyBERT* [240], *YAKE*, or *RAKE* [241]. We use KeyBERT and combine the top three keywords with `OR` logic to ensure broader results. The keyword search itself is done with commercial API.

9.4.3 Relevance Classification

The content of a page p is embedded using SBERT [45]. In the following we use pages and their embeddings interchangeably. SBERT provides dense vector representations of text, capturing semantic similarity between pages. Compared to more traditional methods like TF-IDF or word embeddings like Word2Vec, SBERT excels at understanding contextual nuances and generates sentence-level embeddings that are more suitable for similarity comparisons in semantic search tasks. This allows for a contextual representation of a page's content, even in cases where keyword-based approaches may fail to capture deeper meanings. Once the page content is embedded, the cosine similarity between the embedded vectors of two pages p, q is calculated: Cosine similarity measures how similar the vectors are, independent of their magnitude, making it ideal for comparing the relative semantics of web pages. For two pages p, q and a set of pages Q we define $sim(p, q) = \text{cos\_sim}(p, q)$ and $sim(p, Q) = \max([\text{cos\_sim}(p, q) | \; q \in Q])$. This way, two page p, q are semantically related $p \sim q \iff sim(p, q) \geq \theta$, for a given relevance threshold θ. With this, we define a page p and set of pages Q being semantically related $p \sim Q \iff sim(p, Q) \geq \theta$.

9.4.4 Multi-Armed Bandit

Multi-armed bandit algorithms are crucial for balancing exploration and exploitation in focused crawling. We provide a comparison of MAB algorithms in Table 9.2.

Based on Table 9.2, UCB1 is the best fit for our setting based on its strong performance in stable environments[2], balancing exploration and exploitation with logarithmic regret. It efficiently uses stable domain knowledge to optimize long-term performance without complex tuning. Though slower to adapt to changes, its effectiveness in Eq. consistent environments makes it ideal for forward crawling tasks. (9.1) provides the reward function $r \in \mathbb{R}_{\geq 0}$ for a step $pull(p)$ for page p, the set of seeds S, a domain weight δ, and the number of discovered $|domains|$.

$$r(p) = \max(\delta \; |domains(pull(p))| + \sum_{p' \, in \, pull(p)} p' \sim S, 0) \tag{9.1}$$

[2] Stable in terms of information it seeks, rather than the environment of the web as a whole, which is, as discussed, a dynamic environment.

Table 9.2 Comparison of multi-armed bandit algorithms

Algorithm	Best For	Strengths	Weaknesses
Epsilon-Greedy	Static or semi-static environments	Simple to implement; tunable exploration	Inefficient in complex, large environments
UCB1	Stable environments, long-term performance	Logarithmic regret, efficient exploration-exploitation	Slow to adapt to dynamic changes
EXP3	Adversarial or highly dynamic environments	Robust to non-stationary rewards, no explicit exploration	Over-explores in well-structured settings
EXP3-IX	Non-stationary but less adversarial environments	Better balance of exploration, robust	More complex to implement, requires tuning
Sliding-Window UCB	Non-stationary, fast-changing environments	Adapts to time-varying rewards	Needs careful window size tuning

9.5 Evaluation

The evaluation was performed on a machine with an Intel Core i7-8565U processor and 16GB of RAM. A relevance threshold of 0.6 was used to classify relevant pages. The seed threshold was set to 0.8, as a higher threshold of 0.9 resulted in nearly identical outcomes, which was undesirable for exploration. A blacklist excluded social media platforms like X and YouTube, thread aggregators, non-HTTP protocols, and file types such as images, documents, and program-specific formats. The number of steps was set to 500 and 2000 to allow sufficient exploration actions. The seed consisted of 17 pages, covering security news and in-depth reports.

9.5.1 Evaluation

We evaluate our system based on the combinations of the different search actions, where **B** indicates backward search, **F** indicates forward search, and **K** indicates keyword search.

TC_BFK Balance broad coverage, deep relevance, and keyword-driven exploration.
TC_BF Follows forward links and analyzes backlinks to discover related pages, providing both broad and targeted exploration.
TC_FK Follows forward links and uses keyword searches to expand coverage, capturing more diverse content.
TC_BK Analyzes backlinks and performs keyword searches for targeted exploration and broader discovery.
TC_F Follows forward links from seed pages, offering extensive coverage but with potential inefficiencies.
TC_B Analyzes backlinks to uncover related pages not directly linked to the seed.
TC_K Search solely based on keywords of given pages.

Table 9.3 presents the results of the evaluation. For 500 steps we reached a maximum harvest rate of $\sim$ 23.86% with **TC_F**, followed by **TC_FK** with $\sim$ 20.6%. The worst result achieved **TC_K** with a harvest rate of just 2.7%, which, overall, performed the worst of all combinations. When backlink search was present in the crawl, it performed the best, expect for **TC_BFK**. In terms of domains, **TC_BK** was able to spread the search the widest with 873 searched domains with 132 of them being relevant, despite crawling only 2344 pages (including the seed). Those included multiple security overview and dataset pages[3], as well as multiple unknown security news pages.

During the 2 000 step test, we saw much more balanced harvest rates over the used methods (near to or over 20%, except **TC_B** and **TC_K**) with forward search gathering the most relevant pages on average. The top harvest rate was achieved by **TC_BK** with 25.14%. Keyword search alone still performed the worst but combining it with backlink search achieved the highest harvest rate and gathered the highest amount of relevant domains. In total, nearly a quarter of all crawled pages are of different domains (with **TC_BK**) showing the impact of the keyword search through search engines, and we identified in total 270 relevant domains (with **TC_BF**).

[3] E.g, https://malpedia.caad.fkie.fraunhofer.de and https://github.com/CyberMonitor/APT_CyberCriminal_Campagin_Collections.

Table 9.3 Result of the evaluation with a maximum of 500 and 2000 steps. $|S|$, $|P|$, and $|P{+}|$ show the number of steps (if not the maximum), the number of total crawled pages, and number of relevant pages, of which are $|Seed|$ new seeds, HR the harvest rate [in %], and $\lceil\sim\rceil$ the maximum similarity of identified pages. $|Dom|$ and $|Dom{+}|$ indicate how many domains and relevant domains are identified, respectively. TM shows the top method during this run with $\overline{\text{TM}}$ as the average similarity of this method. Best results are highlighted with light background color, and the worst darker background color

Method	S	$\|P\|$	$\|P+\|$	$\|Seed\|$	HR	$\lceil\sim\rceil$	$\|Dom\|$	$\|Dom+\|$	TM	$\overline{\text{TM}}$
					500					
TC_BFK	128	6 199	387	24	6.24	0.9	751	96	F	0.11
TC_BF	11	405	32	6	7.9	0.9	145	24	B	0.17
TC_FK		9 110	1 877	30	20.6	1	384	23	F	0.25
TC_BK	278	2 344	339	9	14.46	0.9	873	132	B	0.2
TC_F		9 982	2 382	28	23.86	0.87	244	38	F	0.29
TC_B	111	1 313	97	12	7.39	0.9	397	61	B	0.15
TC_K	21	148	4	1	2.7	0.81	105	4	K	0.24
					2 000					
TC_BFK		21 663	4 889	46	22.57	0.92	1 484	189	F	0.38
TC_BF		26 715	4 965	56	18.59	0.92	1 683	270	F	0.29
TC_FK		18 270	4 119	39	22.55	1	809	49	F	0.27
TC_BK		8 175	2 055	19	25.14	0.9	1 965	269	B	0.27
TC_F		21 710	4 992	48	22.99	1	372	52	F	0.27
TC_B	117	1 290	102	12	7.91	0.9	387	61	B	0.15
TC_K	22	155	5	1	3.23	0.81	111	5	K	0.24

9.6 Discussion, Limitations & Future Work

THREATCRAWL effectively addresses the challenge of identifying and crawling CTI related sources from the web, which answers our research question "How can CTI related information be identified and crawled from the web" (**RQ**). By utilizing a MAB approach and seed URLs, the system efficiently expands relevant web content, offering a targeted crawling strategy that is well-suited for CERT and SOC personnel. Unlike related crawlers [41, 229], THREATCRAWL focuses on expanding a predefined set of pages, aligning with the current needs and priorities instead of keywords. In terms of efficiency, the system significantly outperforms related work.

While [41] reported a harvest rate of $\sim 9.5\%$, our system achieves a much higher rate of $\sim 25.14\%$, indicating a more effective method for discovering relevant information. THREATCRAWL is able to identify key information sources relevant to the initial search domain and even expand the current seed by over 300% without relaxing its focus (**C1**). It went over the course of nearly 2000 different web domains and identified 270 relevant ones with starting just 17 seeds (**C2**).

Limitations Despite these positive results, THREATCRAWL faces some key limitations. Firstly, it should be noted that the related page search functionality, as described in [40], is no longer available due to the removal of this feature by search engines[4]. This removal has resulted in the loss of a key capability for searching across a wide range of sources. Second, no comparisons with others [41] was conducted, preventing a broader benchmarking of performance besides the one stated in the references directly. Runtimes exceeding 2000 steps were not evaluated, so we did not observe how the crawler behaved over time or whether it reaches saturation with certain search actions. Tipping points for the crawling parameters are not evaluated either, *e.g.*, using to lax thresholds, using to distinct seeds, or using to few seeds. Finally, embeddings are generated using SBERT rather than larger models like LLaMA or GPT, which could offer improved semantic accuracy but at the cost of higher computational demands and additional privacy considerations.

Future Work Future work could include dynamic adjustment of thresholds based on real-time crawler performance, allowing better adaptability. User feedback from CERT and SOC personnel could also be integrated to guide the system's relevance assessments. Building on that, while using a MAB is definitely a top choice to decide, which actions are more promising in the long run and if there is no information of the page that is used for the current search action. But with crawling this information is just a `GET` away. For example, a very detailed page with very few links could yield better results with keyword or backlink search. Otherwise, if the page is very new, backlink would probably yield worse results than the others. Using graph-based approaches to map and analyze relationships between crawled pages could provide deeper insights into the structure of the CTI landscape. This could be combined with local-sensitive hashing to identify news aggregation platforms and pages that copy others, as well as, the first publisher of information. Other improvements could be a multistep-depth crawling, *i.e.*, if an irrelevant page is reached keep crawling for n steps just to be sure. Such an approach could be combined with URL

[4] Query a search engine with *"related:page_url"* returned related pages, excluding those of the given domain.

classification [49] or adaptable domain blacklisting. While the focus of this paper is the CTI domain, the system should perform well on other domains too, since it is primarily based on the used seed. This aspect needs to be evaluated further.

9.7 Conclusion

CTI information is published in unstructured form on the web, which presents a time-consuming task for CERT and SOC to maintain an up-to-date list of web pages to visit for such information. Our proposed THREATCRAWL system addresses the challenge of identifying previously unknown and relevant CTI sources from the vast amount of unstructured public information available online. By using a MAB approach, it efficiently expands a seed of URLs, making it highly suitable for security personnel who need to automate this time-consuming task, even with a small amount of seed URLs ≤ 20. With a harvest rate of over 25%, THREATCRAWL outperforms prior work, uncovering previously unknown information sources and datasets. However, limitations like short evaluation runtimes and reliance on SBERT leave opportunities for further enhancement. Future work should focus on optimizing search actions, adjusting thresholds dynamically, and leveraging larger models for better accuracy and adaptability.

10 Navigating the Shadows: Evaluating the Dark Web for Cyber Threat Intelligence

Abstract

In today's world, cyber-attacks are becoming more frequent and thus proactive protection against them is becoming more important. Cyber Threat Intelligence (CTI) is a possible solution, as it collects threat information in various information sources and derives stakeholder intelligence to protect one's infrastructure. The current focus of CTI in research is the clear web, but the dark web may contain further information [41]. To further advance protection, this work analyzes the dark web as open-source intelligence (OSINT) data source to complement current CTI information. The underlying assumption is that hackers use the dark web to exchange, develop, and share information and assets. This work aims to understand the structure of the dark web and identify the amount of its openly available CTI related information. We conducted a comprehensive literature review for dark web research and CTI. To follow this up we manually investigated and analyzed 65 dark web forum (DWF), 7 single-vendor shops, and 72 dark web marketplace (DWM). We documented the content and relevance of DWFs and DWMs for CTI, as well as challenges during the extraction and provide mitigations. During our investigation we identified IT security relevant information in both DWF and DWMs, ranging from *malware toolboxes* to *hacking-as-a-service*. One of the most present challenges during our manual analysis were necessary interactions to access information and anti-crawling measures, *i.e.*, CAPTCHAs. This analysis showed 88% of marketplaces and 53% of forums contained relevant data. Our complementary semi-automated

Minor changes regarding language, style, and formatting have been made in this chapter to provide a consistent experience.

P. Kühn, *Proactive Cyber Threat Intelligence*, Technology, Peace and Security | Technologie, Frieden und Sicherheit,
https://doi.org/10.1007/978-3-658-49241-0_10

analysis of 1 186 906 onion addresses indicates, that the necessary interaction makes it difficult to see the dark web as an open, but rather treat it as specialized information source, when clear web information does not suffice.

Original Publication ***Kuehn, P.***, *Wittorf, K., Reuter, C., "Navigating the Shadows: Manual and Semi-Automated Evaluation of the Dark Web for Cyber Threat Intelligence,"* IEEE Access, *vol. 12, pp. 118 903–118 922, 2024.* https://doi.org/10.1109/ACCESS.2024.3448247.

| IF 3.4[1]

10.1 Introduction

Information technology becomes more ubiquitous with every day, and with it, the risks of cyberattacks and cybercrime grows. In 2023, the estimated costs of global cybercrime have reached 8.15 trillion dollars [242], driven by hackers who continuously develop novel techniques to attack IT systems and personnel [2, 5, 6, 32, 41, 243, 244]. Those attacking techniques are shared on the web by attackers in their communities. This includes vulnerabilities, hacking tools, malware, and hacking services [244]. The dark web, in particular, plays a crucial role in this process.

The web comprises three layers: the surface web, accessible via ordinary browsers and search engines, the deep web, which includes non-indexed pages that are usually only accessible through authentication processes, *e.g.*, internal networks, and the dark web, which can provides encryption and anonymity through protocols [80]. Both, surface- and deep web provide no anonymity, making these parts of the web less attractive to criminal hackers [93, 245–247]. Only their availability is still counting towards those services [247]. The dark web, which provides criminals a space for their activities provided in Hidden Service (HS), is a crucial source of intelligence for understanding and mitigating cyber threats [243, 246, 248].

Hence, it is necessary to establish new protection measures and derive information about potential attacks, which is one of the key goals of CTI. The idea in this work is to provide an overview and analysis of current information sources in the dark web, whether they can be used to obtain such information, as the benefits of open information sources cannot be denied [249]. Hence, potentially relevant information can be found on the dark web. In addition, the dark web may be an early

[1] Updated 5. Februar 2025.

source of information, as attacks are planned here so that countermeasures can be developed as early as possible based on this information.

Goal The focus in this work is on gaining knowledge of the benefits of the dark web for IT security. In doing so, the The Onion Router (Tor) network is considered as it is basically a gateway to the dark web. The Tor network is the best known and largest network.

During our investigation, we plan to answer the following research questions:

RQ1 **How can IT security-related information, such as vulnerabilities and exploits, be identified on the dark web?**

RQ2 **What challenges exist in extracting and analyzing open dark web data?**

RQ3 **To what extent can the dark web be utilized for CTI?**

RQ4 **What is the nature and volume of IT security information on the dark web, and how does it contribute to CTI landscape?**

This work makes several contributions to CTI and the dark web research. First, it provides a *manual analysis of forums and marketplaces* provides a new perspective for automated CTI research and is complemented by a complementary *semi-automated analysis of HS* (**C1**). The other contribution *outlines current challenges for automatically gathering open information on the dark web*, with a focus on overcoming CAPTCHAs (**C2**).

The paper is structured as follows: Sect. 10.5 reviews related work, followed by the methodology (cf. Sect. 10.3). Our analysis is presented in Sect. 10.4. We discuss the results in Sect. 10.5 and conclude our paper in Sect. 10.6 with future work directions.

10.2 Related Work

This section provides an overview of previous work on crawling and research of the dark web in general, followed by CTI and vulnerability databases in combination with the dark web. The review process is conducted using commonly known digital libraries[2] and a snowball sampling process afterward. We conclude this section with a research gap.

[2] IEEExplore, ACM Digital Library, arXiv, Google Scholar, and Semantic Research.

Automated extraction of information from the web relies on web crawlers. This is no different for the dark web, except the underlying protocol to connect to this part of the web. [250] presents an overview of current research in dark web crawling. It outlines several currently used methods to mitigate hurdles, *e.g.*, semi-automation via human interactions. We outline additional information of current research in Table 10.1. The present work on the other hand aims for a qualitative approach based on a manual analysis combined with a semi-automated analysis, rather than (semi-)automated crawling approach and computational classification.

Table 10.1 Related work on crawling: Empty fields are not clearly outlined in the paper. Source states if either the surface web is used (○), the dark web is used (●), or both (◐). Focus states if it is a focused (✓) or unfocused (✗). The frequency (Freq.) is either one-time (·), periodically ( ), or incremental (↗). The queue states whether a breath-first search queue (↔), depth-first search queue (↓), or a combination of both (↔↓) is used. The architecture is given as either centralized (⊙) or parallel (‖). CAPTCHAs are either handled manually (⊥), automatically (⊤), or are ignored or not necessary to deal with (✗)

	Goal	Source	Focus	Freq.	Queue	Arch	CAPTCHA	Tools
[257]	IoT-related CTI	◐	✓	·	↓	⊙	⊥	ACHE, elasticsearch
[279]	Extremist forums	●	✓	↗	↔↓	‖	⊥	
[264]	Use free tools	●	✗	·		⊙	✗	AppleScript
[280]	Categorize HS	●	✗	·		‖	✗	
[270]	Databases	○	✓	↗		‖	⊥	Selenium, Bash
[56]	Hacking info	◐	✓					
[277]	CTI	●	✓					
[281]	Products	●	✗			‖	⊤	Python
[282]	Adaptive crawler	◐	✓		↔	⊙		Apache Nutch
[55]	CTI	●		↗				
[283]	State of HSs	●	✗			‖		Docker, Selenium, Cloud
[265]	CTI	●	✗	↗	↓	⊙		BeautifulSoup, keras
[284]	Various	●					⊥	
[41]	IoT-based CTI	◐	✓	↗	↔	‖		Selenium, ACHE
[285]	Marketplaces	●						Scrapy, Privoxy
[286]	Agora	●	✗	↗			⊥	Docker, PHP, cURL
[287]	Drugs	●	✗	·	↔↓		⊥	Selenium, AWS
[288]	HS	●	✗					Scrapy, Selenium
[289]	Anomaly	●			↔		⊥	
[263]	Criminal & illicit content	●	✓			‖		Privoxy
[290]	Captcha Breaking	●				‖	⊤	Selenium
[291]	Drugs	●					⊥	
[292]	Measurement of DWM size	●	✗		↓	‖	⊥	

Others delve into a variety of dark web themes, including drug trafficking, hacker activities, and terrorism. The Darknet Identification, Collection, Evaluation with Ethics (DICE-E) framework [251] emphasizes ethical considerations, guiding researchers in safe and ethical dark web exploration. [252] presents problems and further mitigations and recommendations. All those studies not only contribute to academic knowledge but also aid in developing strategies to counter cybercrime and terrorism effectively. Compared to them, we focus mainly on CTI information. [253] present another survey of current dark web research.

Researchers are exploring various aspects of the dark web, including asset exchange (such as tools or exploits) for cyberattacks [67]. Research methods encompass supervised machine learning (ML) for cyber-security relevance classification of text, with Naive Bayes (NB), Random Forest (RF), Support Vector Machine (SVM), and Logistic Regression (LR) utilized [56, 254–257]. Multinomial classification categorizes forum posts into nine classes for cyber threat extraction [255]. Unsupervised ML, such as clustering, identifies topics, and product categories [258, 259]. Researchers also detect new threats using terms associated with malware, ransomware, data leaks, or vulnerabilities [244, 260]. Additionally, predictive models aim to forecast cyberattacks, treated as binary classification problems [261, 262]. Social Network Analysiss (SNAs) offer deep insights into dark web communities, enabling the analysis of social connections and actor identification [246, 259, 263, 264]. Source code analysis classifies attachments and tutorials for programming languages and exploit types [67, 265–267]. Studies vary in data collection methods, some utilizing existing data [56, 254, 258, 268, 269], while others involve active crawling [56, 67, 246, 257, 260, 265, 270–272]. The research scope ranges from forums to marketplaces, with some covering multiple platforms concurrently [55, 244, 246, 251, 254, 267, 273, 274]. These results are still in line with the results identified by [94]. In comparison to related work, our approach is rather holistic in nature. Our objective is to identify all information that is somehow related to cybersecurity and accompanying threats.

Current research on how dark web information can be used to enhance information in vulnerability databases, remains limited [56, 261, 262, 275]. One objective of this research is to predict cyber threats, often framed as binary classification problems, considering whether vulnerabilities are exploited [261, 262, 272, 275, 276]. Common Vulnerabilties and Exposures (CVE) data, crucial for cybersecurity, serves as a key parameter in this analysis [56, 244, 255, 277]. Some studies employ explicit CVE mentions in posts, using techniques from SNA and ML like RF to correlate hacking activities with vulnerabilities and real-world cyber incidents [262, 276]. Others expand the scope by considering posts mentioning vulnerabilities or extracting Common Platform Enumeration (CPE) data for a multi-label classification approach, utilizing models like SVM and RF [56]. The [41] introduces an integrated framework that monitors data sources, including National Vulnerability Database (NVD) and Twitter, considering CVE as a critical parameter. It aids in identifying, collecting, analyzing, extracting, integrating, and sharing CTI [41]. Additionally, research efforts strive to link exploits to vulnerabilities by extracting exploit and vulnerability names, *e.g.*, a vulnerability identifier (CVE-2014-0160) or their given name (here *Heartbleed*), from dark web forums, utilizing advanced techniques such as Bi-long short-term memory (LSTM) and attention mechanisms [278]. Moreover, a novel vulnerability severity metric is developed, combining vulnerability information with data from linked hacker exploits [278]. As stated, we aim to present a holistic view of available CTI information for the selected marketplaces and forums, based on a manual, rather than automatic analysis.

Research Gap The benefits of OSINT sources for information gathering can not be denied, and the dark web needs to be analyzed further in this regard [249]. Research on the dark web as a potentially relevant source for CTI is researched in various papers. However, current research looks for specific information (tool, vulnerabilities, exploits, or posts) rather than trying to provide a thorough overview of information, which is one of the key aspects of CTI [41, 56, 67, 293]. For data gathering, researchers usually use page-specific crawlers with manual interventions, *e.g.*, manually solving Completely Automated Public Turing tests to tell Computers and

Table 10.2 Overview of CTI from the dark web based on the goal of the research, used tools, and source of the dataset. The sources is a combination of boards (B), websites (P), marketplaces (M), chats (C), or social media (S). Unfilled fields are not shown in the corresponding paper

	Goal	Tools	Source
[257]	CTI relevance ranking	Gensim, SVM	BMP
[254]	Classify hacking related info	Supervised learning	M
[294]	Classify data breach	Supervised learning	B
[56]	Analysis tool of DWF	Supervised learning	BM
[246]	Board analysis		B
[258]	Hacker DWF analysis	SVM, LDA	B
[259]	Comprehend dealer behavior on DWM	K-Means, SNA	M
[295]	CTI with exploit focus		B
[296]	DWM and DWF information link	SNA, gephi	BM
[272]	Current threat campaigns	SVM, SGNS, RF	

(Continued)

Table 10.2 (Continued)

	Goal	Tools	Source
[297]	Correlate threats and dark web discussions	RegEx, LP, PFR	BM
[269]	Threat prediction	LSTM, NER, RF	B
[274]	CTI	CL-LSTM	M
[260]	CTI	Elasticsearch	BMS
[273]	Key players, social connections	SNA	B
[266]	Hacker tools, their key players and source code, CTI	SVM, LDA, SNA	B
[251]	Analyze hacker community		BCM
[255]	Detect CTI in DWF	CNN, SVM, DT, BoW, n-gram	B
[267]	Analyze malware	LSTM, RNN	B
[55]	Integrate various sources in CTI platform	RegEx	BCM
[265]	Visualize cyber threats	RNN, LSTM	B
[41]	Collect and analyze IoT CTI	CNN, NLP, ML, NER	BMP
[67]	Collect and analyze exploit source code	DTL-EL	BM
[278]	Collect exploit source code and link to CVE	LSTM, DSSM	B
[244]	Predict cyber threats	MLP	M
[261]	Predict cyber threats	SNA, ML, RF	B
[262]	Predict incoming cyber threats for organizations	Supervised learning	B
[276]	Predict exploit usage	Supervised learning	BM
[68]	Early warnings of cyber threats	RegEx	B
[275]	Correlate cyber activities and current threats		B
[289]	Automated threat detection	LSTM	M

Humans Aparts (CAPTCHAs) and provide the session tokens to the crawler [291]. Others [268] use pre-compiled datasets to assess the availability if CTI information, without assessing the gathering process as a whole. A thorough analysis of the suitability of the dark web for current CTI information gathering using solely open information and assessing the problems for (partially) automated crawlers and possible mitigations is not done as far as we know (Table 10.2).

10.3 Methodology

This section describes the used methodology to conduct our research. First we specify the methodology for the manual investigations part, followed by the semi-automatic analysis. In addition, we stated the ethical considerations during our research to protect ourselves from harmful content. The whole research process is conducted by two information security researches. In total, we manually analyze 65 DWFs, 7 single-vendor shops, and 72 DWMs. The semi-automatic analysis is based on a provided database snapshot timestamped at 26 June 2022.

10.3.1 Investigation

In Sect. 10.2 a comprehensive literature review has already been conducted, which forms the information basis. Based on these papers, various marketplaces and forums are examined manually. Hereby, we aim for a qualitative analysis, rather than a quantitative [298]. Implementing and maintaining handwritten crawlers is possible during the course of a study, they quickly become out of date afterward, due to the dynamics of the dark web, hindering the reproducibility of the whole study. Moreover, while automated approaches seem to dominate, it is not an uncommon way to approach this field [298]. In order to gain as much insight as possible into the dark web for IT security, it is necessary to manually visit as many HS as possible. For this, preventive security measures are necessary to ensure the safe browsing experience and download of the web pages. Further, the scan is run in a virtualized environment so that an infection cannot spread to other resources. At the same time, JavaScript is turned off by default in the browser so that unwanted JavaScript code is not executed [251]. To get potentially relevant HS entry points into the dark web, we use various clearnet websites[3] and filter them for security related links. Those pages tag links based on their provided content. As the work by [299] suggest, we filter for *digital* as well as *security*. Following this, we contacted these entry points on three separate occasions: November 2022, shortly afterward, and February 2023, aiming to detect possible downtimes. In the fully manual investigation, the search is limited to DWF, single-vendor shops, and DWM, which are accessed after collection. In addition, the Distributed Denial-of-Service (DDoS) mechanism is documented, and user accounts are created if necessary. This is often a prerequisite for exploring the site, especially for DWFs and DWMs. A user account rarely requires a valid e-mail address on the dark web. Such email addresses are created in advance for the

---

[3] https://torhoo.com/, https://tor.fish/, https://darktrain.express/, and https://darkurl.site/

registration process. During the inspection process for the login and registration process, the protection mechanism and the requirements for entering the site, such as an invitation or payment, are documented. For all emerging CAPTCHAs a screenshot is taken to understand the limitation and variations of the mechanisms and to support future research in this area. To access additional information from the page, the operator may require further actions, such as solving programming challenges or achieving a higher level of trust through interactions. The documented information should provide insight into the extent to which the information is publicly available and the hurdles that must be overcome for automatic extraction. For information extraction, we navigate through the (page) menu and document, which categories and subcategories are security relevant and how many entries they contain. Hereby, we follow the keyword list in Fig. 10.1. This list also includes keywords like *phishing* which are part of two worlds, namely, CTI, as it might focus one's organization, and cybercrime. Such an overlap might occur in the other areas of the present work as well. After that, various CTI-relevant categories are browsed, in order to gain impressions of IT security events. For specific information, *e.g.*, to locate CVE mentions, we use website searches when available. Other information that we look out for are: offerings of hacking services, scamming, or social engineering (incl. phishing). Likewise, FAQ and profile settings are visited to gain more information if necessary. In DWF, the first page of threads is visited in relevant categories and the first page of posts is visited for relevant threads. When browsing we document the IT security relevance, occurrences of specific topics, special features, and CVE mentions. The relevance is documented as either *high*, *medium*, or *low*. In DWM, we visit the first page of each relevant category. We browse relevant products, categories, CVE mentions, and document their IT security relevance. To achieve traceability of the research, we store all visited HTML pages. Additionally, the assessment of the IT security relevance is made by two researchers, proficient in the field of IT security, and differences during the assessment are discussed in case of a tie. In doing so, a domain is assigned to one of the three relevance classes *high*, *medium*, or *low* based on the number, variety, and strength of the relevant information. The assessment includes parameters like (i) availability of the service, (ii) number of posts or products related to IT security, (iii) number of posts or products overall, or (iv) policies regarding hacking or IT security. If a domain contains no signification amount of information related to IT security (≤ 10 or ≤ 0.1 % posts/products), it is classified as *low*. If it contains only a few posts or products (> 10 or > 0.1 %) and allows hacking related communication, it is classified as *middle*. And if a page contains many posts, either in absolute or relative amount (> 500 or > 30 %) it is classified as *high*. It should be noted that not all products/posts are reviewed, but only the first page of each (sub)category, thread, or board. A full analysis is considered part of automated procedures and automated processing due to the amount of data involved and leads to different problems, *e.g.*, automatic CAPTCHA-breaking, dealing with automatic logins, generating interactions, and solving other challenges.

IT security keyword list

0-Day, Zero-Day, Zeroday, 0day, Exploit, Malware, rootkit, snort, yara, clamav, DDoS, DoS, Denial-of-Service, denial of service, out-of-service, Ransomware, ransom, Hacking, hacked, Vulnerability, vulnerabilities, vuln, vulns, Botnet, bots, c2 bots, c&c, command and control, Attack, SQL Injection, Keylogger, Phishing, trojan, virus, XSS, cross-scripting, compromise

Fig. 10.1 A compilation of essential IT security vocabulary used to identify and classify digital threats, vulnerabilities, and attack vectors in modern computing environments.

10.3.2 Semi-Automatic Analysis

In order to achieve a higher diversity of HS and to integrate different types of HS in this work, a further step is to visit more HS in a semi-automated way. The goal of this step is to get an impression about the IT security related content and its page types of the dark web. The onion addresses are extracted from an existing dark web database generated by the *Panda Projekt*[4] using a keyword list. This database is generated by regular automatic crawls of the Tor network [300–302]. Figure 10.4 visualizes this process.

The non-all-inclusive keyword list includes various IT security-related keywords including their spelling mistake counterparts (*cf.* Fig. 10.2). We extracted 1 186 906 onion addresses this way. During preprocessing, version 2 onion addresses ($\approx$ 562 591) and duplicates ($\approx$ 297 506) are removed, resulting in 5 560 unique base domains containing all remaining addresses, *i.e.*, 5 560 distinct pages to continue analyzing. For the time being, we restrict ourselves to the total of 5 560 base domains for further manual analysis and do not consider subdomains and paths, because of the high number of paths and subdomains and the associated complexity. This seems to be a good abstraction, since on the one hand subdomains should often match the category of the main domain and on the other hand the paths sometimes refer to JavaScript files that do not reflect the content of the page in a meaningful way. After preprocessing the onion addresses, all remaining addresses are visited

[4] https://panda-projekt.de/

automatically to get an overview of their raw and textual content, which is then stored in the database. During automatic crawling, Privoxy[5] is used to connect to the Tor network. CAPTCHAs are not handled during this process.

The onion addresses are not tracked in depth and thus only one level of breadth-first search is performed. We save the raw HTML content and the extracted textual content. By saving the textual content in this way, it is possible to obtain a lot of information about the website without having to visit it manually. In addition, this protects the researchers from visiting pornographic web pages, since they get a textual overview beforehand[6]. Such web pages are removed at this examination level based on a keyword list[7]. All HSs that are inaccessible during this crawl are crawled again after some time, to catch, whether it was only a short outage. Afterward, the content of the page is analyzed. At the same time, duplicates with the same content, but different domain, are removed. All other textual content is manually examined and, if it indicates promising results, the website is visited. The relevance assessment (*high*, *middle*, *low*) follows the same procedure as in the manual analysis.

10.3.3 Ethics

We explicitly provide the ethical viewpoint for the present work based on work by [303–305]. [306] present further information on critical aspect of darknet research[8]. We provide an overview of why we consider our work research on open information. All HSs investigated are publicly accessible on the dark web. If necessary, accounts are created using cost-free dark web email providers, accessible to all. These accounts access public information without interacting with the community or operators, ensuring no disruption or hacking occurs. No posts, purchases, or payments are made, and server operations remain unaffected. This principle follows previous research standards [290, 298, 307]. [307] states: "The data we collected is essentially public. We did have to create an account on Silk Road to access it; but registration is open to anybody who connects to the site. We did not compro-

[5] https://www.privoxy.org/

[6] Other non-CTI-related content, such as drugs or weapons, were not considered harmful by and for the authors.

[7] We compiled a pornographic keyword list based on https://relatedwords.io/pornographic and added additional words which this list missed: porn, masturbat, lesbian, penetration, pussy, rape, sex, erotic, anal, pedo, incest, dick, gangbang, cock, foreskin, naked, glan, vagina.

[8] The darknet serves as underlying infrastructure for the dark web, similar to relation of the internet and the world wide web

mise the site in any way". Furthermore, only data that does not contain personal information and is also publicly accessible, such as product information or discussions in forums, is examined in this work. We do not make the data publicly available, as it may contain personally identifiable information or metadata, which, combined with other data, could result in re-identification of individual.

10.4 Analysis

We present the results of manual analysis of DWFs and DWMs and the semi-automated extraction of information from various onion domains. First we outline the used sources and their benefits for IT security (*cf.* Sect. 10.4.1). These were collected as part of the literature review and manual exploration process. We follow up with IT security relevant information in these sources (*cf.* Sect. 10.4.2). Additionally, we discuss the relevance of HSs. We conclude this section by highlighting challenges of Information Retrieval (IR) in the dark web (*cf.* Sect. 10.4.3). The listed challenges are either part of the manual analysis or the automated crawl.

10.4.1 Information Sources

The Tor network provides so-called HSs, which hide the IP address of the server. There are different types of HSs, *e.g.*, DWM, single-vendor stores, DWF, and chats. Not all HS are illegal or have added value to IT security. From [308] it appears that out of 1000 HSs, approximately 68 % had illegal content.

- **Dark web marketplaces** facilitate product and service trade [56, 296], with specialized markets focusing on distinct products [309]. Drugs are a primary commodity [286, 309], while ≈13% are related to hacking [56, 277]. Users include traders, customers, and administrators [291]. Cryptocurrencies like Bitcoin (BTC), Monero (XMR), or Ethereum (ETH) facilitate transactions [310]. Marketplace data includes product and trader details [55, 285], accessible post-payment [278]. Insights from marketplaces aid in understanding cybercrime markets [311], serving as early warning systems for potential data leaks [278].
- **Vendor shops**, smaller marketplaces, offer limited product variety and insights compared to larger markets [298].
- **Dark web forums** are akin to Clear web forums [277], enabling user communication [56]. Varying in size and topics, they are hierarchical and contain threads classified into categories. Posts contain textual content, hyperlinks, media, and other resources in languages like English, Chinese, and Russian [246, 251, 278].

Rich in text and metadata [55, 246, 296], forums include participant lists and user profiles with registration and last visit timestamps [246]. They offer insights into the identity, motivation, strategies, goals, and tactic, technique, and procedures (TTPs) of hacking individuals [55, 270, 295, 312]. Forums also host relevant CTI-related topics [56, 256, 266, 278, 284, 311].
- **Internet Relay Chat** channels allow real-time plain text message exchange among members [251, 278]. Contents aren't usually archived, requiring active monitoring [55, 251]. Internet-Relay-Chat (IRC) channels offer insights into TTP and outbound threats from hackers [55]. For Cyber Threat Intelligence (CTI), IRCs may be promising, given hackers' freely communicated knowledge and goal sharing due to the lack of archiving in IRC channels [284].

10.4.2 Possible Information and Relevance of Hidden Services

This section is about what information was found in the various information sources and how the relevance ranking relates to the information found.

10.4.2.1 Information Uncovered in the Dark Web

Dark Web Marketplaces In Fig. 10.3 we present the extracted product types on marketplaces and their total occurrences. The following categories are identified:

- **Account Hacker**: Offerings of paid services for hacking accounts.
- **Hacker Services**: Offerings of hacking for cell phones, computers, websites, grade improvements, email/SMS spamming, phishing, and DDoS. These services require interaction and don't disclose target information.
- **Social Media Enhancement Services**: Provide likes, followers, views, comments, or members for platforms like Instagram, YouTube, and Twitter, impacting disinformation campaigns, recruitment, awareness, and advertising.
- **Phishing and Scam Sites**: Offered with guides and tutorials for creating them.
- **Tutorials**: Cover various topics including botnet, phishing, 2FA bypass, social engineering, and penetration testing.
- **Tools and Cracked Software**: Offerings of viruses, trojans, Remote Administration Tool (RAT), ransomware builders, antivirus, anti-detection, MAC address changers, malware removal, and vulnerability scanners. Also, cracked software versions like Adobe Photoshop, WinRAR, and CyberGhost are available.
- **FullZ/Combolists**: Offerings of email lists, password lists, and personal information.

Example for CVE extracted data.
"RedLine is one of the most widely deployed information stealers that can grab Windows credentials, browser information, cryptocurrency wallets, FTP connections, banking data, and other sensitive information from the infected hosts. I want to highlight that there is 2 version of this stealer, Lite and Pro. I am selling the Full Red Line Stealer with the whole available Options, in Lite Version most advanced Options are not active and disabled. so kinda useless. be aware what you are buying. some readings."

REDACTED REFERENCES BY THE AUTHORS CONTAINING THE CVE-ID `cve-2022-1096`.

Not included in Fig. 10.2a are standard password lists or onion link lists due to rarity. Some markets sell invitation codes or links for other markets (Genesis Market and Benumb Market), potentially offering different products. Additionally, data extraction found CVE numbers on 7 marketplaces, often linked to the "Trillium Security MultiSploit Tool", featuring exploit generators for various CVE numbers. One such example is presented in Fig. 10.4. The range of relevant products across the analyzed marketplaces in this paper varies between 0 and 1.7000×10^4. In Figs. 10.2a and 10.5, an estimated upper bound of relevant products is plotted on the x-axis. This upper bound is based on the number of products within all relevant (sub)categories of the marketplace. It's important to note that not all products within these categories may be crucial (false positives), and there might be relevant products in uncounted categories (false negatives), such as misclassified items, though this is less common. Additionally, this data reflects the analysis period, and marketplace product numbers continuously change.

Dark Web Forums Parallel to marketplaces, Fig. 10.3 displays forum topics, post content, and total post counts, showcasing significant differences in content and size among visited forums. Approximately 5 out of 18 forums prohibit hacking or illegal content, rendering them irrelevant to this paper's research. Figure 10.3a outlines forum themes, including marketplaces and forums prohibiting trading. Some forums host marketplaces for product advertisements or requests, like the CryptBB forum offering ransomware, keyloggers, RAT, hacking books, and 0-Days. Others feature account hacking services and DDoS attacks. The second category encompasses valuable sources for learning hacking skills, including programming resources, SQL injection examples, hacking challenges, and hacking books,

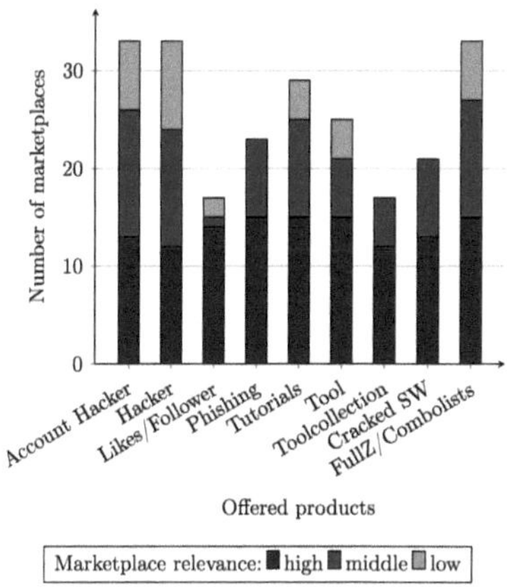

(a) This figure shows the offered product types (with a focus on IT security product) on dark web marketplaces. The relevance of a dark web marketplace is indicated by the tint of the color. The darker the color, the more relevant is a dark web marketplace.

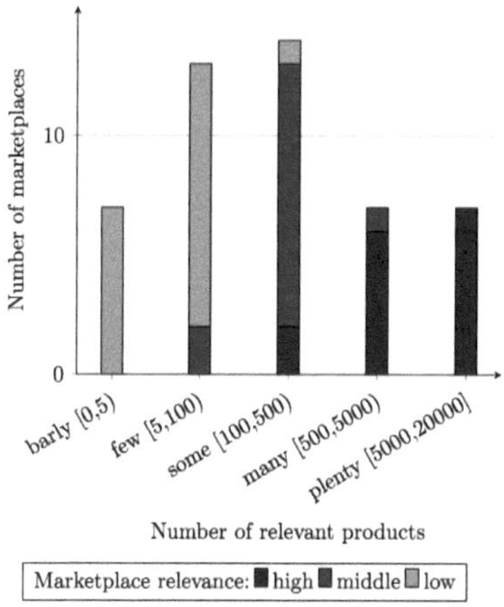

(b) Estimation of the number of relevant products that exist in the marketplaces, divided into intervals intended to reflect the states "there are hardly any or none, few, some, many, very many relevant products in the marketplace".

Fig. 10.2 The left shows what is offered on marketplaces, while the right shows how many products exist on the marketplaces. In both graphs 48 marketplaces are considered

offering the most relevant information. Forums also mention various ransomware, RAT, and hacking tools like phishing tools, stealers, keyloggers, and crypters. Threads in marketplaces contain ransomware source code or Android RAT. The "cracked data/programs" section includes posts disclosing Twitter databases or hacked PayPal accounts, akin to forums where individuals offer hacking services as Hacking as a Service (HaaS). Notably, the bulk of relevant information comes from higher-relevant forums, with significantly fewer contributions from lower-relevant forums, many of which prohibited hacking content (*cf.* Fig. 10.3b). In Fig. 10.3b, the post counts are displayed, albeit with some missing data due to access issues such as forum unavailability, registration problems/restrictions[9], or payment requirements. CVE numbers were discovered in 7 forums, enabling successful CVE searches (*cf.* Fig. 10.7a. These instances feature vulnerability news, security recommendations, and the sale of exploits. Compared to marketplaces, forums more frequently lack IT security-related information. About 47 % of forums and 13 % of

[9] Some forums require challenges to register, *e.g.*, proof of black-hat hacking activities.

marketplaces yield no relevant information, while interesting product types are almost always present on marketplaces.

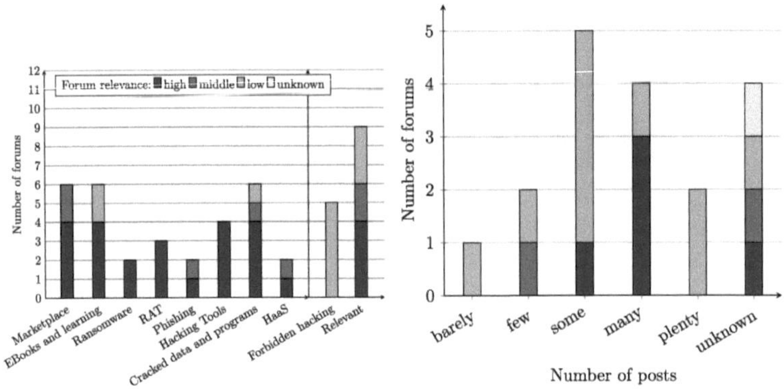

(a) Most relevant information originates from higher relevant forums, with notably less from lower relevant ones. This conclusion was drawn from exploring the initial page of each interesting thread. However, this method might have missed some content, indicating that the figure does not provide a comprehensive view of the forums.

(b) In contrast to the marketplace figure, this count represents total posts, not specifically relevant ones. The study included significantly more low-relevant forums. However, post counts remain unknown when pages were inaccessible or login wasn't possible. *Barely* indicates up to 5×10^2 posts, *few* up to 1×10^3, *some* up to 1×10^4, *many* up to 3×10^4, and *plenty* up to 5×10^6 posts.

Fig. 10.3 Post content and topics are shown on the left, and the total number of posts is shown on the right. Both bar charts show 18 forums

Hidden Services In Fig. 10.4, the investigation process with final classification of 1 186 906 onion addresses is depicted. Due to issues like invalidity, duplicates, subdomains, and paths (step 1), only 5 560 addresses were automatically crawled. Of these, only 1 698 were manually classified (step 3) due to non-reachability, content duplication, and the removal of pornographic material (step 2) (cf. Sect. 10.3.2). However, Step 2 did not eliminate all such pages, leading to their re-inclusion in step 3 (9 %) where content was textually saved for classification. Categorization of Onion domains involved dividing them into seven classes. Many websites, particularly in the single vendor stores category, exhibited similar content and page structure, potentially indicating phishing sites, though these were not deeply examined in Fig. 10.4. A significant proportion (74 %) of manually classified Hidden Services were single vendor shops. Despite low IT security relevance, the availability of hacked money accounts added interest. Another class encompassed hidden wikis, directories, and search engines, not providing direct IT security insights but serving as gateways to relevant sites. The *Miscellaneous* category (7 %) comprises various

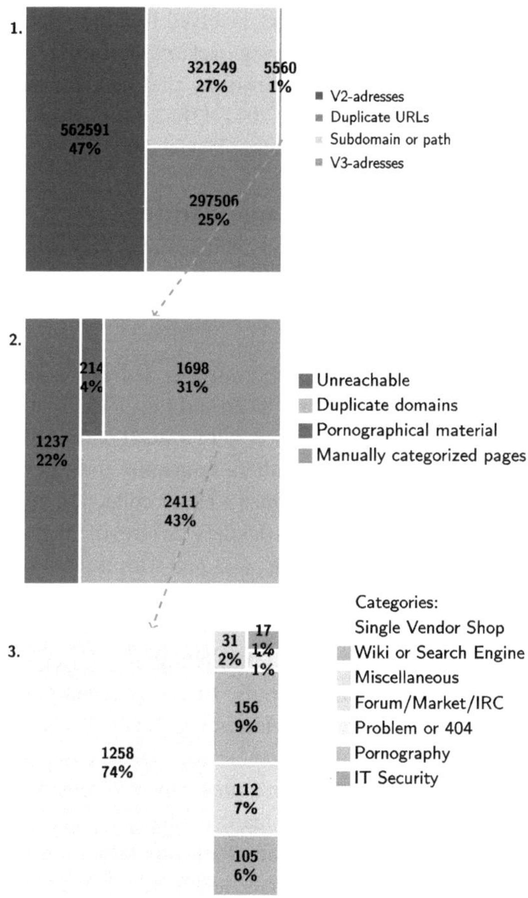

Fig. 10.4 Procedure for examining hidden services. While steps 1 and 2 are automated, the last step is performed manually. In step 2 we look for duplicates based on the base domain. It should be emphasized that only 0.00143% of all given onion addresses have a verifiable, direct link to IT security

hidden services, including political pages, private individual sites, Bitcoin-related information, and a range of tools/services like Bitcoin generators, email providers, escrow, and hosting services. The *Forums, Marketplaces, and IRC* class includes sites covered in manual analysis within the present work and others not found using the entry points in this work. Challenges outlined in Sect. 10.4.3 emerged throughout the analysis, including linguistic barriers and blocked pages taken down by authorities. The most relevant category involved hacking-related pages (1 %). Prominent were single vendor stores offering hacking services lacking information about tactics, techniques, and targets. Noteworthy sites included an exploit builder for CVE numbers CVE-2018-0802, CVE-2017-11882, CVE-2014-1761, and CVE-2012-0158, a RAT, and a database containing leaked data.

10.4.2.2 Relevance Factors Interdependencies

In this study, relevance relies on the quantity and nature of pertinent products/posts, as outlined in the methodology (*cf.* Sect. 10.3). This section delves into the factors affecting relevance using the gathered data.

Marketplace Relevance Marketplace relevance is tied to the number of potentially relevant products, depicted in Fig. 10.2b and Fig. 10.5. Total product counts often don't align with IT security relevance, with a small portion attributed to such products as seen in Fig. 10.5, where outliers frequently indicate restricted offerings. Usually, higher relevance accompanies a larger collection of hacking-related products. However, four outliers in Fig. 10.5 defy this trend: three at slightly lower relevance but with more diverse products, and one with limited offerings despite medium relevance.

Correlation with Product Types and CVE Numbers The variety of product types (*e.g.*, Account Hacker, FullZ, Tutorials, Tool) significantly influences a marketplace's relevance (*cf.* Fig. 10.6). A wider range generally enhances relevance, yet Fig. 10.2b lacks a definitive correlation between offered product types and relevance. Drawing conclusive statements requires a more extensive investigation of additional marketplaces. Low-relevance markets often lack extensive tool collections, possibly due to a scarcity of hacking tools and limited inventories held by traders. Notably, hacking services, tutorials, and combo lists persist as prevailing goods across all marketplaces. Another correlation can be seen based on the finding of a CVE number. In all 7 cases the marketplace relevance was classified as high when a CVE number was found.

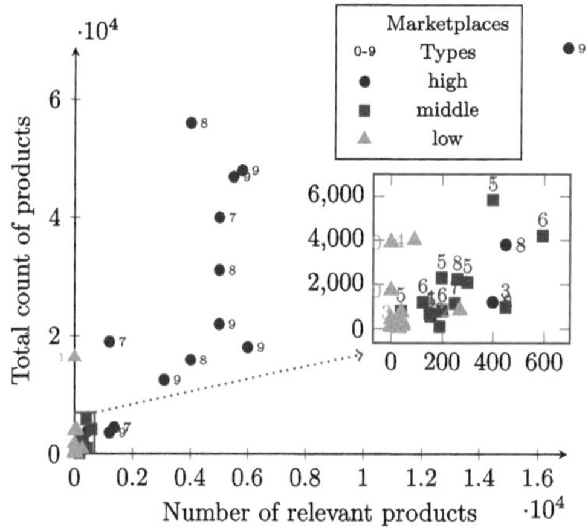

Fig. 10.5 The chart illustrates relevance concerning the number of products available. Some marketplaces ignore the topic hacking, leading to lower relevance despite a high product count of up to 9 products. In this graph, outliers can be justified by the diversity of product types offered, as indicated by the label *Types*. The relevance is classified as *high*, *middle*, or *low*

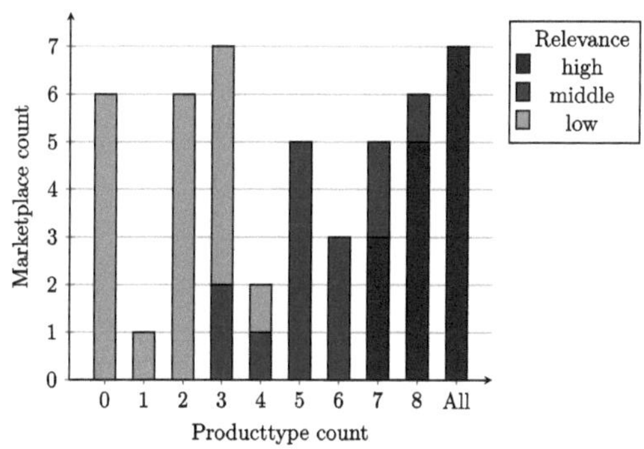

Fig. 10.6 This graph shows the relevance in relation to the variability of the product types. The product types are listed in Fig. 10.2a and include, *e.g.*, FullZ, tools, tutorials or hacking services. It can be clearly seen that the number of product types offered has an influence on the relevance of a marketplace

Forum Relevance

The relevance of forums heavily relies on the diversity of topics covered in the posts. Highly relevant forums tend to encompass a broader range of IT security-related themes (*cf.* Fig. 10.7b). However, relevance doesn't directly correlate with the discovery of a CVE number, as shown in Fig. 10.7a. Among the 4 forums where CVE numbers were present, noticeable differences exist: high relevance forums (here 2) and low relevance forums (here 2). Further investigation based on the data might confirm varying occurrences across relevant forums. High relevance forums contained an offer of an exploit related to a CVE number, while low relevance forums contain links to clear web vulnerability descriptions. Relevance correlates with potential relevant posts and the ratio of potential relevant threads, as shown in Fig. 10.8. Yet, direct correlation with the total count isn't clear due to limited data. Forums surpassing 500 relevant threads and 1000 relevant posts tend to demonstrate high relevance, while those below indicate medium or low relevance.

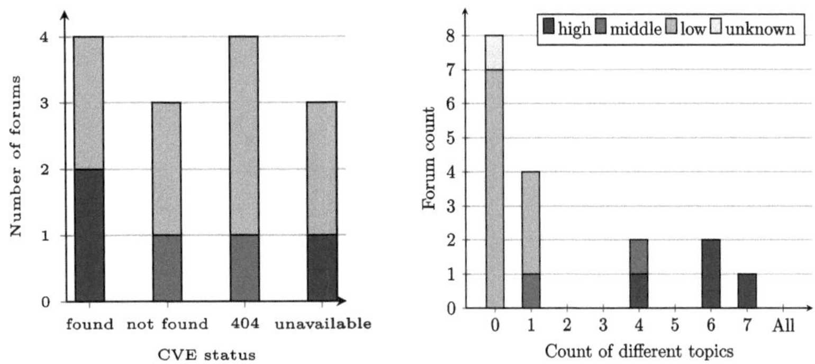

(a) Show the number of forums, which contain CVE information and their corresponding relevance, indicated by their color. The more relevant, the darker the color.

(b) Show the number of forums plotted against the count of different topics classified as security relevant. The colors of each bar depict the relevance, in terms of security, of the examined forum. The more relevant, the darker the color.

Fig. 10.7 The relevance of a forum depends on the number of different topics in the forum

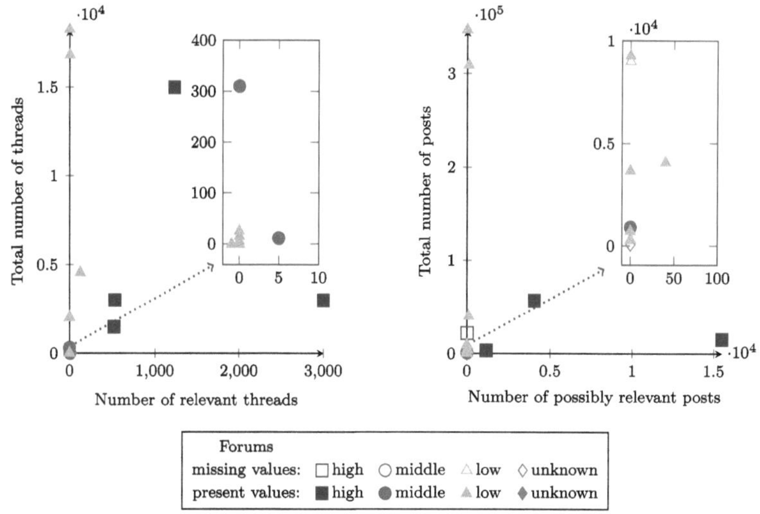

Fig. 10.8 The forum's relevance hinges on the count of relevant threads (left) or posts (right). Graphs illustrate that exceeding 500 relevant threads or over 1000 relevant posts denotes high relevance, while falling below these thresholds indicates medium or low relevance

10.4.3 Challenges of Information Gathering and How to Deal With Them

Information collection involves manual or automated methods, addressing two main aspects: identifying onion addresses (dark web domain addresses) and documenting the content [293]. Several challenges arise in this process, requiring resolution, acceptance, or mitigation, as elaborated in the subsequent discussion (Table 10.3).

Fast Pace The dynamic nature of the dark web presents significant challenges in information collection [56, 254, 277, 282, 313]. Transient onion addresses, frequently undergoing changes or deactivation, exacerbate these challenges. These addresses often face seizure or go offline, further complicating consistent access. In addition, the presence of multiple addresses for single sites and intermittent downtime due to server maintenance or issues add to the complexity. The evolving structure of hidden services complicates automated crawling, necessitating regular updates to crawling scripts and increasing maintenance efforts [264].

Table 10.3 Summary of challenges and their solutions or mitigations

Challenge	Description	Mitigations
Fast Pace	Hidden Services are dynamic and availability is sometimes instable.	Re-crawl multiple times and keep crawlers up-to-date to get access or confirm unavailability.
Speed and Linking	The bandwidth of Tor is smaller and inter-page linking is relatively poor.	Use clear web Tor directories, hidden directories, or search engines to spread the crawl seed.
Language Diversity	The examined web pages were mainly written in English or Russian.	Use a translation engine or specialized language models for analysis.
Access Restrictions	Multiple DWFs and DWMs have access restrictions to open information.	Use CAPTCHA breakers or plan with necessary interactions.
Scalability	Queuing systems of HSs prolong crawling and file downloads may contain malware	Prepare malware scanners and distributed crawling.
Inconsistencies & Heterogeneity	Sites follow no strict schemata to allow easy crawling.	Specialized crawlers tailored to possibly only one site must be written.
Research/Tools	Datasets for the dark web are not published and tools are quickly outdated.	The entry fee to get going is higher than in the clear web.
Anti-Crawl Mechanisms	CAPTCHAs are a massive problem as they await everywhere.	Manual action might be necessary or further research into CAPTCHA-breakers.

Speed and Linking The Tor network, burdened by its encryption, decryption, and routing processes, operates at a slower pace compared to the conventional clearnet which presents a notable disadvantage for information extraction [251, 283]. The absence of a centralized repository for onion addresses, coupled with the reluctance of many HS operators to increase their visibility, impedes the possibility of a comprehensive analysis of all HS. A substantial portion of the content remains non-linked and, therefore, inaccessible, adding to the complexity of data acquisition [246, 314]. Furthermore, redundant and looping links on the dark web pose additional challenges in automatic data extraction processes, necessitating careful consideration in their handling [56].

Language Diversity In the HS of the Tor network, the most common languages we encountered where English and Russian. For automatic analysis, translation engines can be used, although the quality of these cannot currently be compared to a native human speaker [246, 279]. In addition to the variety of languages, abbreviations and code words are used [69]. Those code words make it hard to understand the actual meaning [293].

Access Restrictions
Many marketplaces and forums have implemented access restrictions, as detailed in (cf. Table 10.4). Sites often require an account to access information, including posts or product lists, necessitating registration. In certain instances, forums require payment (in $\approx 28\%$ of the cases) or an invitation during registration. Sites demanding registration fees or vouches present legal and ethical dilemmas for researchers due to potential support of criminal activities [246, 251, 295, 298] (*cf.* Sect. 10.3.3), leading to non-interaction in this study. Approximately $\approx 36\%$ of accessible, understandable forums offer additional information or rights for a fee ranging from $5 to $250 (*cf.* Fig.10.9). Marketplaces, similar to those in the Clearnet, require payment for products, granting access to detailed product information. Dark web marketplaces often impose a security fee on sellers, refundable or bypassable under specific conditions, varying from $50 to $1 000 (*cf.* Fig. 10.9). This seller fee explains the higher average prices, being a deposit in some cases. Fig 10.9 reveals that the highest fees are associated with more relevant marketplaces/forums. In marketplaces, higher fee tiers correlate with a larger product range, whereas in forums, especially those requiring an entry fee, the exact number of posts remains unknown. In a subset of forums, specifically $\approx 24\%$ of those analyzed, additional interaction mechanisms are employed. These interactions may include solving a challenge, writing posts, or responding to specific content. The *CryptBB* forum exemplifies a combination of these mechanisms, requiring users to solve a challenge and make three public forum posts to access more information. Moreover, responding to the original post can unveil additional hidden content. A similar approach is adopted in other dark web forums, where interaction with hidden content is facilitated. DWM frequently employ trust levels as an access control mechanism. Users can advance to higher trust levels through purchases, ratings, and secure account settings. In AlphaBay market, it was observed that new registrants at trust level 0 could not view all products; for instance, only two out of four products listed in a category were visible. [307] noted in her research that Silk Road hosted secret offers, not publicly accessible but available only through URLs distributed via out-of-band mechanisms like private messages (*e.g.*between seller and buyer). Moreover, traders could operate in a 'secret mode,' with their products and information page not directly linked by

Table 10.4 This table presents an overview of all accessible forums and marketplaces that are at least partially in English, totaling 25 forums and 48 marketplaces. The table aims to indicate the availability of information without requiring a login. It also offers insights into the frequency of successful logins, the data necessary for registration, and the factors hindering registration. These hindrances include issues like server errors, blocked IP addresses, JavaScript requirements, or suspended registrations

Properties	Options	DNF #	DNF %	DNM #	DNM %
Openly available	No	11	≈ 44%	26	≈ 54%
	Some	6	≈ 24%	2	≈ 4%
	All	9	≈ 36%	20	≈ 42%
Registration information	Username & Password	17	100%	46	100%
	E-Mail	11	≈ 65%	5 ≈	11%
	E-Mail Validation	1	≈ 6%	0	0%
	PGP-Key	3	≈ 18%	1	≈ 2%
	PIN	0	≈ 0%	23	≈ 50%
	Login/Personal passphrase	0	≈ 0%	5	≈ 11%
	Nickname	0	≈ 0%	9	≈ 20%
Registration hurdles	Successful	10	≈ 40%	42	≈ 88%
	Not available	3	≈ 12%	1	≈ 2%
	Payment/ Invitation	7	≈ 28%	0	≈ 0%
	Problem	4 ≈	16%	4	≈ 8%
	Offline (not tried)	1	≈ 4%	1	≈ 2%
Actions for information	Admin Approval	3	≈ 12%	On marketplaces, it is common to move up trust levels through interactions (purchases)	
	Payment	9	≈ 36%		
	Invitation	3	≈ 12%		
	Interaction	6	≈ 24%		

SilkRoad [307]. The AlphaBay case suggests the presence of either secret products or products visible only after attaining a certain trust level.

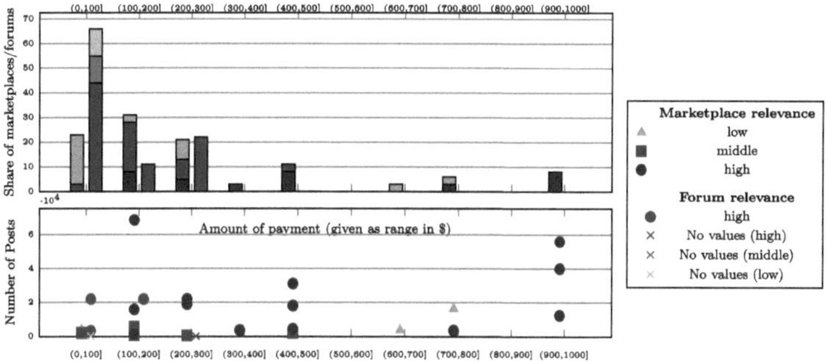

Fig. 10.9 This graphic depicts the potential costs involved, encompassing all forums (9) and marketplaces (40) that require a payment. In forums, payments are predominantly for access, additional information, or extended rights, whereas in marketplaces, payments are generally for obtaining trader status, in addition to product purchases. Note that the costs of products themselves are not included in this graphic

Scalability The Tor network hosts numerous websites ($\approx$ 750000 onion addresses[10]), with individual sites often featuring many subpages. Among the HS analyzed in this study, the largest marketplace offered around 68 000 products, while the largest forum hosted about 350 00 posts. For practical reasons a prioritized approach is essential. Large-scale data extraction requires automated crawlers capable of handling unstructured and inaccessible data in a scalable manner. Distributed crawling is one effective method [285]. Consideration of the memory load for downloaded content and efficient resource utilization is critical. Given the dynamic nature of content, especially on forums, high throughput is necessary to maintain data coverage and freshness. However, excessive requests to the same domain can overwhelm a website, inadvertently triggering a DDoS attack, an outcome to be avoided [246, 263, 279, 285]. To prevent overload, many marketplaces (approximately 27 % analyzed) have implemented a queue system (*cf.* Fig. 10.10), more frequently than forums. This security mechanism, independent of a marketplace's IT security relevance or size, ensures availability. The queue mandates a waiting period before entry, prolonging both manual and automated crawling times.

[10] Date 2023-12-22, https://metrics.torproject.org/

Crawling multimedia files significantly increases total crawling time and doubles the average file size compared to HTML files alone. Forums typically contain a mix of content types, from static files like HTML and PDF to dynamic text and multimedia data. To avoid the additional load and risks associated with multimedia files (e.g., malware or illegal content), they can be detected and annotated rather than downloaded [270].

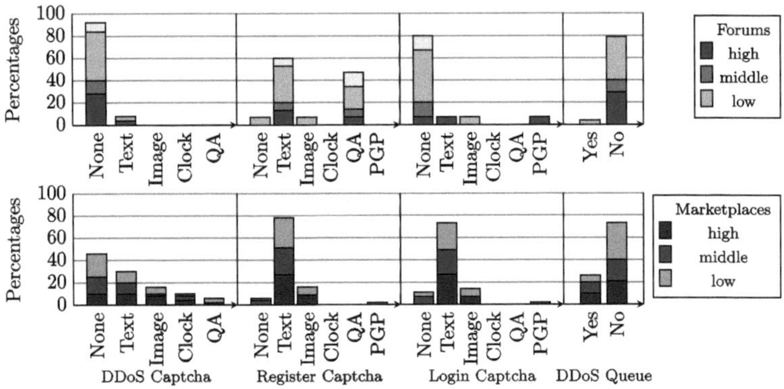

Fig. 10.10 It can be seen that forums have implemented far fewer DDoS captchas than markets. At the same time, marketplaces have more difficult DDoS captchas. In the forums examined, captchas often only occur when registering and not when logging in, while both are often implemented on marketplaces. Furthermore, it can be seen that a DDoS queue is implemented far more frequently on marketplaces than on forums

Inconsistencies & Heterogeneity The dark web's diversity of forums and marketplaces necessitates tailored understanding and customization for automated crawling [251]. Units of measurement vary across sites; for example, seller deposits are in Euros on Cocorico Market, Bitcoin (BTC) on Babylon Market, Monero (XMR) on DarkOrbit Market, and US-Dollar (USD) on others. Conversions, subject to fluctuating rates, are momentary solutions; automated processing must also account for varied spellings of the same values (*e.g.*, "50 USD" vs "$50") [309]. Posts and products are often inconsistently categorized, misplaced in wrong or multiple categories, or found as empty or duplicate listings. Users may also find category assignments unclear, further complicating data analysis [291, 315, 316].

Research/Tools A comprehensive and in-depth analysis of the dark web remains largely underdeveloped. Consequently, there is a scarcity of supportive tools for dark web investigation, and those that exist are often not publicly available, primarily due to concerns over their dual-use potential. For instance, [270] withheld their dataset from publication to prevent misuse. Furthermore, the dark web's rapid evolution renders datasets quickly outdated and obsolete, often covering only a limited number of websites [293, 317].

Anti-Crawl Mechanisms Information extraction from the dark web is further complicated by anti-crawling mechanisms and DDoS prevention strategies, mainly implemented in CAPTCHAs. Some websites have systems to detect excessive requests from the same Tor identity, leading to blocking of that identity, necessitating a change of identity for continued access. To circumvent these mechanisms, automated crawlers can mimic human behavior, such as slowing crawl speeds or introducing random delays between page requests [270]. However, the challenge persists even in manual investigations, requiring the ability to automatically detect and reset the identity upon detection. Additionally, websites may employ CAPTCHA challenges or redirect to a DDoS-CAPTCHA page after a certain number of page visits, as noted by [287] in the context of many marketplaces. These mechanisms extend the execution time for both manual and automated crawling and necessitate additional considerations in the crawling script or the possibility of recall due to incorrect data display [55, 279]. [315] also highlight an anti-crawling technique that randomizes the HTML/CSS of login pages to further impede automated data collection. The most significant anti-crawling challenge encountered in this study is the use of CAPTCHA. According to [318, 319], CAPTCHA can be classified into four types: text-based, image-based, video-based, and audio-based. In this research, CAPTCHA were encountered in various situations, including i) DDoS prevention, ii) login processes, iii) registration procedures, iv) post writing, v) handling of numerous emergent requests, and vi) marketplace browsing.

Figure 10.10 presents the frequencies of DDoS, registration, and login CAPTCHA. Notably absent were video- and audio-based CAPTCHAs. Figure 10.11 presents CAPTCHA examples, we encountered during our analysis, ranging from clock CAPTCHAs (*cf* Fig. 10.11a) to chess-like CAPTCHAs (*cf* Fig. 10.11d). It was observed that forums were less likely to implement CAPTCHA mechanisms compared to marketplaces, and the complexity of these mechanisms was also generally lower in forums. This disparity could be attributed to the higher proportion of low-relevance forums studied compared to low-relevance marketplaces. Marketplaces also have a higher likelihood of being targeted for DDoS attacks due to their significant role in illicit activities and their potential impact on criminal trafficking.

The types and difficulty levels of the individual CAPTCHA encountered during this research are detailed, providing valuable insights for future studies on CAPTCHA complexity. It was observed that CAPTCHA appeared more frequently during registration processes in forums, in contrast to marketplaces where CAPTCHA were commonly used at both registration and login stages. The tendency to employ CAPTCHA mechanisms varied with the relevance of the HS; higher-relevance sites were more inclined to use them. This could be partly due to the larger size of the marketplaces, as they might aim to enhance their security measures. In forums, the likelihood of CAPTCHA implementation increased with relevance, possibly reflecting the advanced IT security knowledge of the operators. The content focus within forums also differed based on their relevance; low-relevance forums generally covered a broader range of topics and frequently restricted hacking-related discussions, while high-relevance forums often included hacking-specific categories. Consequently, the need for robust security measures and the risk of unauthorized access were greater in forums of higher relevance.

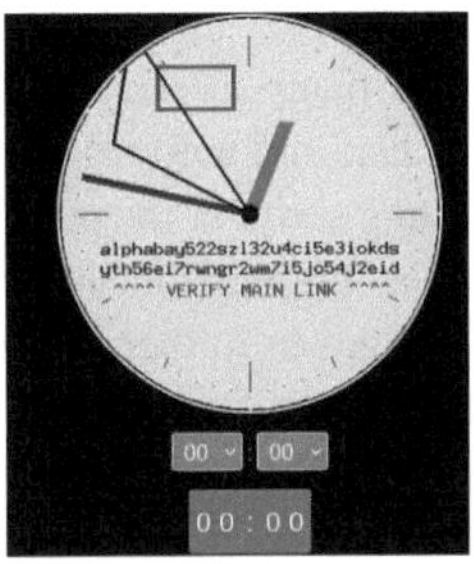

(a) A clock CAPTCHA for DDoS protection from the *AlphaBay* market.

(b) A puzzle CAPTCHA for login protection encountered on *Royal Market.*

(c) A puzzle CAPTCHA for registration protection encountered on *ArcheTyp*.

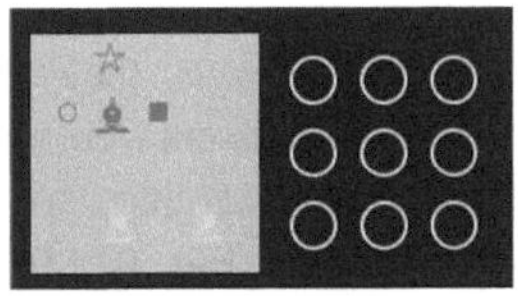

(d) A chess-like CAPTCHA for DDoS protection encountered on *ViceCity*.

Fig. 10.11 Example CAPTCHAs encountered during the study

For automated crawling of CAPTCHA-protected pages, the CAPTCHA must be either solved, bypassed, or the pages ignored [280]. There are three primary methods for tackling CAPTCHA [320]: i) halting the crawling process at a CAPTCHA and requiring manual resolution [270, 284, 286, 289, 309], ii) utilizing external services that employ human solvers for CAPTCHA, such as *Anti-Captcha*[11] and *2Captcha*[12] [315, 321], and iii) developing automated CAPTCHA solvers to circumvent service costs [290, 318, 320, 322]. Figure 10.12 and Fig. 10.13 display the challenges encountered during data extraction from DWM and DWF, in the order experienced during a visit. These paper/darkweb/figures clarify the necessity of solving CAPTCHA to access IT security information. The depicted visit process for a HS encompasses all HS that are accessible and visible without JavaScript and have not been seized. A potential initial barrier is the DDoS-CAPTCHA encountered upon page entry. If this CAPTCHA exists and remains unsolved, no IT security information can be accessed. The availability of information upon entering the page is then assessed. If all information is readily available, no additional efforts are

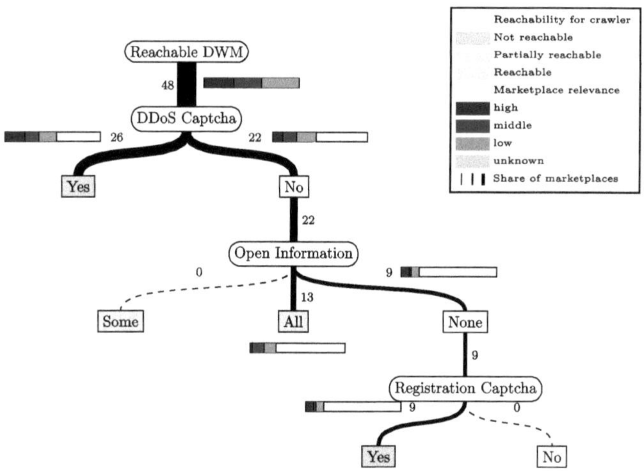

Fig. 10.12 This figure illustrates the historical pattern of DWM visits. It depicts whether relevant IT security information can be extracted from DWM without the need to solve CAPTCHA. The figure effectively shows the sequence of a typical DWM visit. It reveals that in most marketplaces, solving a CAPTCHA is a prerequisite for accessing IT security-related information

[11] https://anti-captcha.com/de

[12] https://2captcha.com/

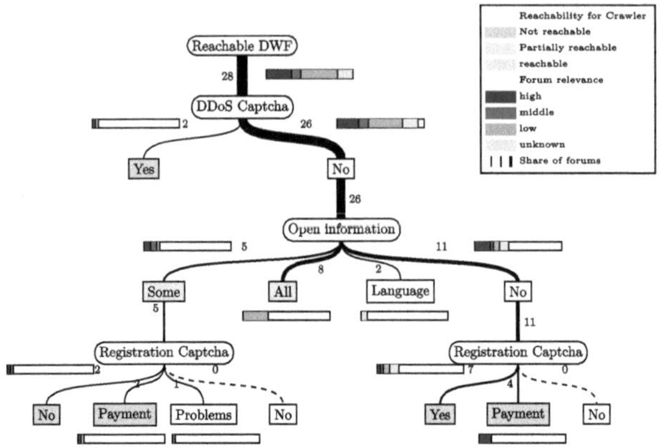

Fig. 10.13 This figure illustrates the history of DWF visits. It demonstrates, based on the progression of a DWF visit, whether relevant information can be extracted from DWF without the need to solve CAPTCHA. The figure indicates that complete information extraction from a forum is typically possible only in low-relevance forums

required. However, if only partial or no information is available, registration, possibly involving another CAPTCHA mechanism, is needed. This ensures restricted information disclosure without CAPTCHA resolution. Subsequently, solving a login CAPTCHA might be required.

In the marketplace analysis presented in Fig. 10.12, 48 marketplaces were examined from which 13 medium and low relevance marketplaces could be accessed without resolving any CAPTCHA. All necessary information can be accessed without solving any CAPTCHA from 8 out of 28 forums, although these forums typically contain limited IT security-relevant information (*cf.* Fig. 10.13). Partial information can be extracted from 5 forums, which vary in their relevance.

10.5 Discussion

The discussion in Sect. 10.5.1 initiates with an inquiry into the methodologies for uncovering IT security-related information on the dark web. This involves a comparative analysis of manual versus automated crawling techniques. Key to this comparison is an examination of the challenges associated with each extraction method. The subsequent segment of the discussion evaluates the dark web's potential as an open information resource. In this context, research questions and are explored in greater depth. This section concludes with the limitations encountered in this study as well as future work (*cf.* Sect. 10.5.3). It outlines the various decisions in the research that imposed constraints. Furthermore, the rationale behind these limitations and their consequent implications are thoroughly discussed.

10.5.1 Manual vs. Automated Crawling

We identified, that IT security-relevant information can be extracted from the dark web through manual or automated crawling methods, which answers **RQ1**. Each approach comes with its own set of pros and cons as identified by the several problems outlined in Sect. 10.4.3. Manual extraction, often criticized for being infeasible and time-consuming, is inefficient and prone to errors [258, 323]. It becomes impractical when dealing with large volumes of data. Additionally, manual extraction may be undesirable due to personal protection reasons, such as avoiding exposure to pornographic content. The process is significantly slower for humans, especially in cases requiring frequent revisits to the same page for monitoring purposes. Ensuring traceability, reproducibility, and facilitating analysis during manual extraction also adds to the time burden. Errors during manual extraction are inevitable due to human factors such as lack of concentration. However, manual extraction is indispensable in certain scenarios. Understanding the structure of forums and marketplaces is crucial for accurate data storage, necessitating manual inspection of the page and analysis of the Document Object Model (DOM) element to adjust the page parser accordingly (*cf.* Sect. 10.4.3). Manual crawling is also essential for gaining a comprehensive understanding of the subject and obtaining valuable insights. As highlighted in Sect. 10.4.3, automated crawling in the dark web is fraught with challenges, answering **RQ2**. A significant obstacle for automated crawling is overcoming anti-crawling mechanisms, as evident in Fig. 10.13 and Fig. 10.12. Many forums and marketplaces necessitate solving a CAPTCHA to access comprehensive information.

Currently, researchers often resort to manually solving CAPTCHAs, necessitating continuous human involvement. Only a limited number of studies have focused on the automated resolution of these CAPTCHAs [290, 318, 322]. Therefore, human intervention or a fully manual crawl is required to gather all pertinent information.

10.5.2 Dark Web as Open Information Source

The primary objective of this study is to utilize the dark web as a OSINT source for IT security related information, as outlined in Sect. 10.1. Section 10.2 presents previous research using the dark web for IT security information is reviewed. This includes extracting new cyber threats, analyzing trends, examining relationships between hidden services, classifying attachments, predicting cyber threats, and generating alerts, answering **RQ3**.

This research substantiates the notion that the dark web is a valuable OSINT resource, demonstrating various use cases. For instance, [260] focuses on generating cyber threat warnings using Twitter and the dark web. The findings, illustrated in Table 10.5 and Fig. 10.14, reveal a lower frequency of warnings based on dark web posts compared to Twitter. Their data sources included 200 dark web and deep web hacking forums and Twitter posts from 69 international security researchers and analysts.

During this study, relevant information was found on marketplaces, forums, and other HSs. The investigation revealed a scarcity of IT security-related information: approximately 88% of marketplaces and 53% of forums contained relevant data. The findings encompass hacking tools, service offerings, cracked software, and leaked

Table 10.5 Cyber threats for which alerts were generated in the paper by [260] and related tweets and dark web mentions

			Mentions	
Discovered Word	Threat type	Warnings	Twitter/X	Dark Web
mirai	Weak Spot	94	537	85
teamxrat	Ransomware	13	30	0
luabot	Trojans	12	27	0
cryptoluck	Ransomware	12	29	0
clixsense	data breach	12	26	5
gooligan	Malware	9	26	0
usbee	Malware	9	18	0
adultfriendfinder	Data breach	9	23	0
starhub	Data breach	9	82	0
badepilogue	Malware	8	21	0
evony	Data breach	8	25	0

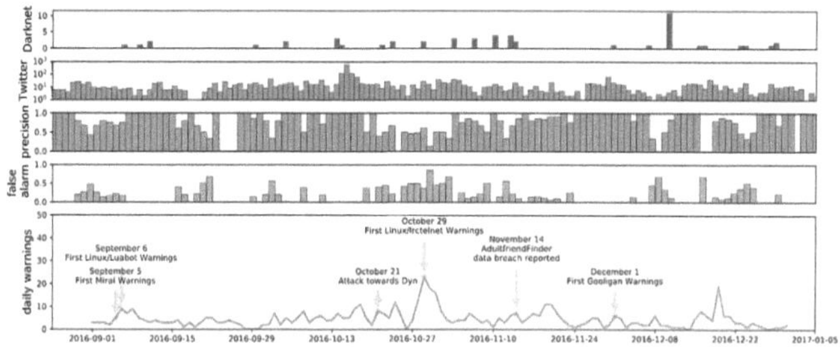

Fig. 10.14 Number of daily generated warnings and the number of related tweets and dark web posts. (Source: [260])

data (*cf.* Sect. 10.4.2.1 and **RQ4**). However, a significant amount of information remained inaccessible due to hidden content, payment requirements, or memberships (*cf.* Sect. 10.4.3). Thus, more information likely remains concealed on the dark web. The semi-automatic analysis indicated that only 0.00143% of all analyzed onion addresses had a direct, identifiable link to hacking, amounting to 17 HSs. [260] suggests that the clear web may yield more information and is potentially easier to access, the cost and effort of using the dark web as an information source must be taken into consideration, which is inline with [324]. It states that Twitter is primarily useful to advertise services and comment on cyber threats rather than patch prioritization. However, this inference must be approached with caution, as the clear web was not evaluated as an information source in this study, thus precluding a direct comparison. Moreover, the quantity of information in the clear web might be outweight by quality in the dark web.

10.5.3 Limitations and Future Work

Initially, the research was confined to the Tor network within the dark web. The list of DWFs and DWMs scrutinized was not exhaustive, comprising only those found in public directories. Despite consulting multiple directories, it cannot be assumed that these directories encompass all existing sites. Due to the fact, that the dark web is a highly dynamic domain, with domains disappearing once in a while, our results can only be seen as a snapshot, which is backed up by findings in related work [41, 56, 260]. In addition, every research, especially manual investigations, inherits some sort of bias of the researchers, which we aimed to mitigate by our chosen and clearly described and strictly applied methodology (*cf.* Sect. 10.3). While the semi-automated analysis identified some additional forums and marketplaces, these were not subsequently examined in detail. Due to complexity constraints, the study did not delve into all posts and products on the forums and marketplaces, focusing instead on just the first page of each. Further, a semi-automated analysis, using a handwritten crawler might have provided additional insights. However, such an attempt would likely produce an overwhelming amount of data to analyze, for our scope and was out of scope for the present work.

The absence of direct interaction also represents a limitation of this research. In summary, there was no interaction with traders. Additionally, the study did not involve responding to hidden content, challenges, or posts in forums, nor were any payments made for memberships or access to information.

A significant challenge in automatic data extraction is overcoming anti-crawling measures, particularly CAPTCHAs. For efficient and fully automated extraction, the development of a robust CAPTCHA-breaker, capable of adapting to the dynamic nature of the dark web, is essential. The current state of the CAPTCHA variants used can be instrumental in future developments. These limitations imply that the scope of information captured in this research is not comprehensive, thereby limiting its ability to represent the full spectrum of IT security information available on the dark web. Another limitation pertains to the exclusion of information from non-German and non-English Hidden Services. Had there been proficiency in languages such as Russian or French, or if automatic translation tools were employed, a broader range of information might have been available for analysis.

Future Work The manual analysis in this study did not encompass the complete extraction and analysis of information from forums and marketplaces, mainly due to the involved efforts. Follow-up studies could monitor open dark web information sources for more IT security-related information tackling and mitigating the identified challenges in an automated manner, including the identification of key users and connections via social network analysis (SNA), and analysis of shared attachments, and source code could yield a more comprehensive and current overview. Forums focusing on hacking and programming are of particular interest in this context. An automated and thorough analysis of IT security-related content in these forums could offer significant insights. Moreover, leveraging a larger and more accurate dataset of the dark web, automated analyses utilizing ML could be conducted, providing additional valuable CTI information.

In the semi-automated analysis phase, onion addresses were initially extracted from a database using keywords. This approach presents upward potential with improved domain filtering from the original data. Various ML/Natural Language Processing (NLP) techniques could be employed to enhance the results. While the results of the present work indicate, that complementing CTI with automated dark web intelligence might be too costly, addressing the question of which information source—dark web or clear web—is more promising, a direct comparative study of these two sources could provide valuable insights. Future research could build upon the literature review and the comprehensive analysis of IT security-related content conducted in this study. The insights gained from this work offer a foundational overview of prior research and a current understanding of the dark web, serving as a basis for all future CTI research focused on the dark web. With a potential availability of Quantum Computing (QC) and Post-Quantum Cryptography (PQC) the current dark web may tremble, as it is based on public key cryptography, which can be broken with [325].

10.6 Conclusion

The dark web is another open information source for CTI [41, 56, 67, 293]. This work provides an overview of existing CTI research and potential CTI-relevant information on the dark web, serving as a foundation for future researchers in designing their investigations. A key goal of this research was to analyze the information available on the dark web and assess its relevance to IT security. The distinction between this work and existing literature lies in the methodology employed and the focus of the information presented [41, 56, 67, 293]. Achieved through manual and semi-automated IR, this study identified that proactive CTI in the form of exploits, malware, and their stakeholders can be sourced from the dark web, whereas others mainly pick one specific aspect to look out for, *e.g.*, Internet of Things (IoT)-based CTI [41]. Leaked company data and cracked software, indicative of vulnerabilities and reactive information, can help protect against further data loss. Contrary to expectations, few visible TTPs or attack targets were discovered in forums. Techniques were mainly found in tutorials and e-books, with ransomware, keyloggers, other malware, and hacking literature being significant finds. Hacking as a service and DDoS as a service offerings were noted, relevant for general security but less so for effective countermeasures unless attribution is involved. Manual investigation of 65 dark web forums, 7 single-vendor shops, and 73 dark web marketplaces revealed pertinent CTI information, while the semi-automated analysis of numerous Onion domains was less fruitful. Despite keyword-based domain extraction, the majority of HS were unrelated to hacking, highlighting the low percentage of IT security-related HS among all HS. The study identified several challenges in analysis and crawling, such as the transient nature and inaccessibility of content, the speed of the Tor network, language diversity, inconsistent categorizations, and scalability issues. Access restrictions like interactions or payments also limit the research scope. Notably, hidden content in forums, accessible via post responses, and account creation pose significant hurdles. Future research will grapple with anti-crawling measures, particularly CAPTCHA at various stages of HS access. This research illuminates new potential CTI information sources. Considering the extraction challenges and information obtained from the dark web versus the clear web, as shown in [260]'s work, the clear web be the primary information source. It can be complemented by dark web collection if (i) current CTI capabilities are not saturated, (ii) clear web data only hints at deeper threats, or (iii) the information is so niche and sensitive that its discussion might be illegal, as dark web collection is more complex. However, valuable insights can still be derived from the dark web, useful for both reactive and proactive purposes. Given the escalating cyber threats, the dark web's relevance as an information source is likely to grow in the future.

OVANA: An Approach to Analyze and Improve the Information Quality of Vulnerability Databases

11

Abstract

Vulnerability databases are one of the main information sources for IT security experts. Hence, the quality of their information is of utmost importance for anyone working in this area. Previous work has shown that machine-readable information is either missing, incorrect, or inconsistent with other data sources. In this paper, we introduce a system called *Overt Vulnerability source ANAlysis (OVANA)*, which analyzes the information quality of vulnerability databases utilizing state-of-the-art machine learning (ML) and Natural Language Processing (NLP) techniques, searches the free-form description for relevant information missing from structured fields, and updates it accordingly. Our paper exemplifies that on the National Vulnerability Database, showing that OVANA is able to improve the information quality by 51.23% based on the indicators of accuracy, completeness, and uniqueness. Moreover, we present information which should be incorporated into the structured fields to increase the uniqueness of vulnerability entries and improve the discriminability of different vulnerability entries. The identified information from OVANA enables a more targeted vulnerability search and provides guidance for IT security experts in finding relevant information in vulnerability descriptions for severity assessment.

Minor changes regarding language, style, and formatting have been made in this chapter to provide a consistent experience.

Supplementary Information The online version contains supplementary material available at https://doi.org/10.1007/978-3-658-49241-0_11.

P. Kühn, *Proactive Cyber Threat Intelligence*, Technology, Peace and Security | Technologie, Frieden und Sicherheit,
https://doi.org/10.1007/978-3-658-49241-0_11

Original Publication ***Kuehn, P.***, *Bayer, M., Wendelborn, M., Reuter, C., "OVANA: An Approach to Analyze and Improve the Information Quality of Vulnerability Databases," in* ARES'21: Proceedings of the 16th International Conference on Availability, Reliability and Security, *ACM, 2021, p. 11.* https://doi.org/10.1145/3465481.3465744.

| ICORE B[1]

11.1 Introduction

Vulnerabilities in IT systems pose a threat to cyberspace. If ignored or overseen, they can lead to disastrous infrastructure failures of states, organizations, and individuals [326], exemplified by the NotPetya outbreak in 2017 [327]. This attack was successful even though the vulnerability was patched three months earlier [327]. Hence, it is imperative to find and report vulnerabilities in existing systems, and make information about them accessible to IT service providers and IT administrators as quickly and detailed as possible [8]. Various vulnerability data sources are available online to support IT security experts in their daily workflow, and thus, secure IT infrastructures and systems.

Leading vulnerability databases are the *Common Vulnerabilties and Exposures (CVE)*[2] and *National Vulnerability Database (NVD)*[3], which have accumulated more than 150k and 160k entries, respectively, as of 2021-06-01. The CVE data feed collects not yet publicly known vulnerability information submitted by researchers, organizations, or individuals, while the NVD migrates it into its own data feed. This information is polished, structured, and enhanced with additional features, *e.g.*, the Common Platform Enumeration (CPE) specifier and a severity score using the Common Vulnerability Scoring System (CVSS)[4]. The CVSS severity score is calculated based on the CVSS vector, which is a key-value representation of different parameters and their values, *e.g.*, the parameter *attack vector* and the corresponding values *network*, *adjacent*, *local*, or *physical*.

However, research has shown that information in both databases is sometimes out-of-sync [39] or offers not enough evidence to allow timely reproduction and patching of the described vulnerability [328], what can be interpreted as a sign of poor information quality (IQ). Similar problems have been shown in other areas of

[1] Updated 5. Februar 2025.

[2] https://cve.mitre.org/

[3] https://nvd.nist.gov

[4] https://www.first.org/cvss/specification-document

IT security, *e.g.*, bug reports [329, 330]. These aspects slow down otherwise urgent action between initial discovery and patch, and create gaps of days, if not weeks or months [331]. The seriousness of this problematic situation increases when one considers the mass of information security experts who face it. Different sources, *e.g.*, vulnerability databases, bug reports, or blog posts, may state different (and possibly even contradictory) facts about a vulnerability, as their knowledge is also based on different sources. To enable security experts to cope with piles of information, it is mandatory to analyze the quality of information available in high-reputation information sources and ideally structure it in a way that machines are able to provide support.

In this paper we analyze the IQ of structured information in the NVD. The NVD is a trusted and tested source of information among IT security experts. Its information is double-checked and research has shown that the CVSS field in the NVD is more credible compared to other databases [332]. However, with more than 30 vulnerabilities disclosed every day, the amount of information seems overwhelming. Therefore, automation to guide and support experts is a promising avenue for research. However, these tools require structured information, *i.e.*, information that allows easy automation of processes. Usually, IQ is considered a subjective metric, depending on the person reviewing it. However, aumann and Rolker [97] have categorized indicators to measure IQ, some of which can be measured objectively. In this paper we seek to answer two research questions:

RQ1 **What are quantifiable criteria for the assessment of IQ in vulnerability databases?**

RQ2 **Which distinct information improves the IQ of the NVD?**

For this purpose, we leverage the structured information in the database entries and the free-form entries. Our metric for IQ incorporates the indicators accuracy, completeness, and uniqueness. Therefore, we compare the vulnerability information of all entries to obtain an overall picture rather than focusing on the details of individual entries. To achieve our goal, several challenges need to be solved: (i) a metric must be defined to measure the IQ within the NVD, (ii) relevant information must be extracted from free-form vulnerability descriptions, and (iii) incorrect information must be rectified and missing information must be added (*e.g.*, the CVSS version 3).

These challenges are tackled with our tool OVANA, which leverages the state-of-the-art in NLP, BERT embeddings [47] and the flair-framework [333] to identify relevant information in free-form vulnerability descriptions and automate the process of IQ analysis. To encourage future work on this project, we provide the source code of Overt Vulnerability source ANAlysis (OVANA)[5].

This paper showcases our contributions with OVANA. On the theoretical side, we provide a metric for measuring the IQ within vulnerability databases **(T1)**. Second, we propose a novel, partially explainable, two-step machine learning pipeline capable of predicting security relevant tags and a CVSS score of NVD entries that can facilitate better vulnerability assessment **(T2)**. As practical contribution, the proposed tool identifies information in the NVD entries that can be used to speed up the search, reproducibility, and patching of the vulnerabilities **(P1)**. Further, the proposed pipeline may be used in the early stages of vulnerability disclosure to assist security professionals in finding an appropriate CVSS value and providing users with vulnerability guidance **(P2)**. Lastly, our tool elucidates the actual assignment of CVSS scores by making the prediction explainable **(P3)**.

The rest of this paper is structured as follows: Sect. 11.2 shows related work and outlines the research gap, Sect. 11.3 introduces a preliminary analysis followed by the concept of OVANA. Section 11.4 shows the evaluation of the ML models used, additional tweaks of the system, and presents the results of the IQ analysis. Section 11.5 discusses contributions and limitations, with a corresponding presentation of future work, while Sect. 11.6 concludes this paper.

11.2 Related Work

This section presents related work, categorized into report and CVSS analysis, followed by the research gap.

11.2.1 Report Analysis

Public vulnerability reports have been examined in a variety of contexts. Nguyen and Massacci [334] are the first researchers to conduct a large scale inconsistency analysis. They identified incorrect version numbers in the NVD entries concerning the Google Chrome browser. Based on their results, Dong et al. [39] analyzed inconsistencies in software versions between vulnerability reports in the NVD and

[5] Published here: https://github.com/PEASEC/OVANA

CVE. They showed that inconsistencies between NVD entries and corresponding CVE entries are highly prevalent and result in missing information in the NVD. Chaparro et al. [329] analyzed vulnerability reports and found missing steps-to-reproduce and expected behavior of the software. Similar work has been done by Mu et al. [328]. They analyzed the reproducibility of crowd-sourced vulnerability reports and chose a manual approach, by checking, whether vulnerabilities in 368 selected reports were reproducible with the information provided. Only 54.9% of the analyzed reports offered enough information to be reproducible, showing a similar result, *i.e.*, a lack of quality in public vulnerability descriptions. Considering the tremendous amount of work that has to be spent on manual quality analysis, the authors stress the need for an automatic collection of the missing information.

Different possible use cases for vulnerability reports have been analyzed. You et al. [169] studied vulnerability descriptions and automatically generated Proof-of-Concept exploits for the Linux kernel using semantic information in CVE descriptions. Their results show that vulnerable functions, variables, and system calls are critical semantic information for automated exploit generation. Schaberreiter et al. [335] presented a methodology for assessing trust in Cyber Threat Intelligence (CTI) sources based on a closed world assumption. They leveraged different IQ indicators, such as timeliness, to compare different CTI sources. Farhang et al. [336] analyzed Android vulnerabilities and their corresponding security bulletin entries from different vendors and discovered structural differences in the reporting on different vendor platforms. Zhang et al. [337] used information provided by the NVD to predict incoming cyber risks by using information about the affected product and the frequency of occurrence of new vulnerabilities for that product. Fan et al. [330] developed a tool to identify valid bug reports using 33 extracted features from bug reports that can be used in various occasions, *e.g.*, bug bounty programs.

Several papers proposed ways to improve the IQ of vulnerability reports. Guo et al. [338] identified missing information about the vulnerability type and trigger of a vulnerability in NVD entries and completed them using deep learning. Anwar et al. [38] analyzed inconsistencies in the structured fields *publication date*, *vendor*, *product names*, *severity score*, and *vulnerability type* in the NVD entries. They rectified these information by crawling the references of the specific entries and used different ML techniques, followed by an analysis of the rectified data.

11.2.2 Vulnerability Severity Assessment

The CVSS was introduced in 2003 with the release of version 1 in 2005 and has since been improved, resulting in the current version 3.1, released in 2019. Over the years, it has been criticized for its lack of transparency [339, 340]. Besides these critiques, the CVSS has been analyzed in different dimensions, ranging from comparing the severity perception of IT security experts with the CVSS scores [341], to correlating the CVSS scores with the number of real-world exploitations [342]. Research by Ruohonen [343] showed an average delay of the CVSS severity assessment in the NVD of one week in recent years. Different proposals bridge this gap with ML methods. Elbaz et al. [224] train a ML model based on the free-form NVD descriptions to predict the CVSS vector. Anwar et al. [38], on the other hand, use ML with the structured information of NVD entries to update CVSSv2 scores into v3 scores.

11.2.3 Research Gap

Vulnerability information sources are mandatory to patch loopholes and consequently secure IT infrastructures. This requires good IQ as timely as possible. Recent work in the field of vulnerability report analysis shows poor report quality [328] and inconsistencies in vulnerability databases [39, 334], which leads to confusion and a delayed patching process. All the works shown either focus on a single IQ indicator or information detail [39, 224, 338], or use solely the structured information to correct information [38]. This urges an automated and universal approach to improve the IQ of official sources such as the NVD, which uses various IQ indicators, corrects erroneous information and predicts empty fields based on all available information, including free-form description. Work that provides such an IQ metric which can be used to compare IQ improvement approaches [38, 39, 224, 338], as well as research that analyzes different indicators of the vulnerability database IQ, is currently still missing. As Elbaz et al. [224] have already indicated, this research needs to be explainable so one understands the decision-making processes of the algorithms.

11.3 Concept

This section shows a preliminary analysis of the NVD to motivate our research. This analysis evolved into the OVANA architecture, followed by an explanation of how we solve the different challenges.

11.3.1 Dataset

In the present work, we focus solely on the NVD. Hence, we use all 146,100 NVD vulnerability entries from 1999 until 2020, timestamped on 2020-05-03, which are available at the NVD as JSON feed[6]. From this dataset, we extract 3,000 vulnerability entries distributed over the years 2013 until 2019, which contain a CVSSv3 vector. Their descriptions are manually labeled with the tags in Table 11.1 (details follow) by two IT security experts. These labeled vulnerability descriptions form the basis for training and testing the Named-Entity Recognition (NER) tagger and CVSS score predictors (*cf.* Sect. 11.3.3). To reduce manual workload and error-proneness, we only labeled the tags from Table 11.1 without annotating the CVSS values, which can be obtained directly from the NVD database. As an example, we have labeled the tag *attack vector*, but not the associated CVSS values *network*, *adjacent*, *local*, and *physical*.

Table 11.1 Labels used in descriptions

Label	Tag	Usage
Software name	SN	[39]
Software version	SV	[39]
Attack Vector	AV	CVSS
Attack complexity	AC	CVSS
Privileges required	PR	CVSS
User interaction	UI	CVSS
Scope	S	CVSS
Confidentiality impact	CI	CVSS
Integrity impact	II	CVSS
Availability Impact	AI	CVSS
Weakness	W	CWE
Vulnerable function	VF	New field
Vulnerable path	VP	New field
other	O	

[6] https://nvd.nist.gov/vuln/data-feeds

11.3.2 Preliminary Analysis

To look deeper into the direction of IQ, we performed a preliminary analysis of the information in the NVD (timestamped 2020-05-03). We examined whether the structured information alone is sufficient to identify vulnerabilities. This is an important step in IT security management, as it enables automated detection of vulnerable IT system in an infrastructure. *E.g.*, if the information in vulnerability reports is too broad, one might find many vulnerabilities unrelated to the specific system resulting in a shut-down of specific components, causing a business process disruption. For this reason, we implemented a small cluster analysis. It clusters the entries in the NVD dataset `nvd` in one year `y` based on their machine readable information: the fields CPE, Common Weakness Enumeration (CWE), and CVSS v3, which are extracted in `reduce_entry`. The clustering algorithm is given in the following Python excerpt[7].

Listing 11.1 Simple clustering algorithm

```
cluster = defaultdict(list)
for v in nvd[y]:
    cluster[reduce_entry(v)].append(v)
```

This method showed large clusters in our dataset, *e.g.*, CVE-2019-2138 to CVE-2019-2165. We further examined our dataset to find the origin of clusters and identified the following problems:

1. Weakness information are missing (*e.g.*, CVE-2019-5620) or use dummy values (*e.g.*, CVE-1999-0003 or CVE-2019-10732).
2. The outdated CVSSv2 standard is used (*e.g.*, CVE-2017-18538).
3. Information in the free-form description is not fully utilized in the structured information (*e.g.*, in CVE-2019-9956 or CVE-2019-1010101).
4. Some entries miss CPE information (*e.g.*, CVE-2017-0235) and in general the CPE is rarely used at its full potential (*e.g.*, CVE-2019-8010 misses the vulnerable version).
5. Libraries are accounted as part of the product they are used in, even if there are matching CPE classes and should be standalone products (*e.g.*, in CVE-2019-9232).

[7] References defaultdict https://https://docs.python.org/3/library/collections.html?highlight=defaultdict#collections.defaultdict

Over the course of this paper we focus on problem 2 and 3, while we leave the rest for future work. Given these concrete problems, we wrote a solver that uses regular expressions to identify path information and CWE classes within descriptions of NVD entries. We use the identified information to enrich the corresponding NVD entries to reduce the previously mentioned problem of having too broad information in NVD entries. This simple method reduced the overall average cluster size from $\approx$ 1.22 to $\approx$ 1.16, *i.e.*, by $\approx$ 0.061 (*cf.* Fig. 11.1). Figure 11.2 depicts the improvements in cluster sizes of this approach (excluding singletons). The first bar of each year shows the clusters of the original data, while the second bar shows the clusters of the enriched data. The gradient color is based on the clusters size, *i.e.*, clusters

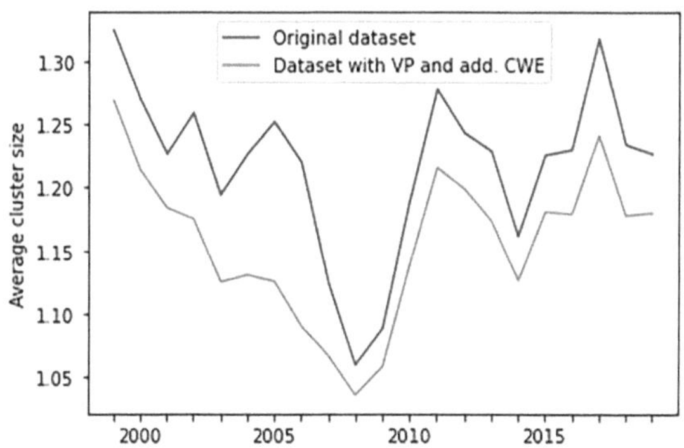

Fig. 11.1 Average cluster sizes per year

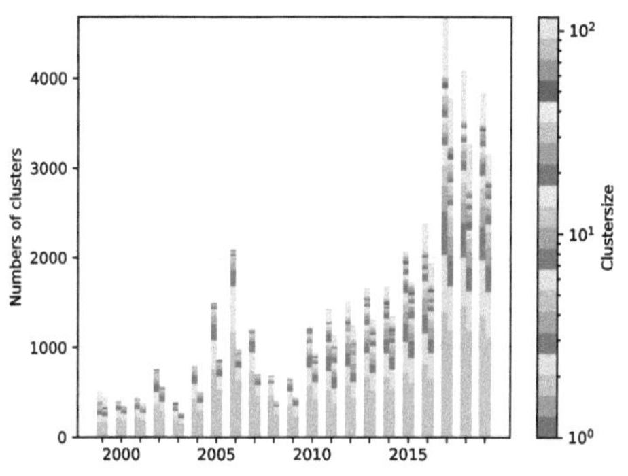

Fig. 11.2 Clusters by year excluding singletons: the first bar of each year shows the clusters of the original data and the second bar the clusters of the enriched data

in the same size range share the same color. One can see that our approach split clusters into smaller ones (depicted by the decreased average cluster size), while some clusters remain quite large. Based on these results, we looked further into possible solutions with sophisticated methods to solve the problems.

11.3.3 OVANA Architecture

OVANA is composed of different modules: an IQ module, a NER module with a tagger for each tag, and a CVSS value predictor module with a predictor for each CVSS parameter, while the IQ module is used to benchmark the IQ. The IQ module benchmarks the IQ score of the input data based on the IQ indicators used, while the NER and CVSS modules are used to increase the IQ value of the this data. Specifically, the NER module is able to identify important named entities in the description of NVD entries. The CVSS value predictor module consists of multiple deep-learning models to predict the value of a specific CVSS key based on the identified named entities. The output of the NER module and the CVSS value predictor are then implemented into the vulnerability dataset by either correcting inconsistencies (in case of the CVSS value) or adding identified core information (in case of the vulnerable function (VF) and vulnerable path (VP)). Figure 11.3 depicts the pipeline as described.

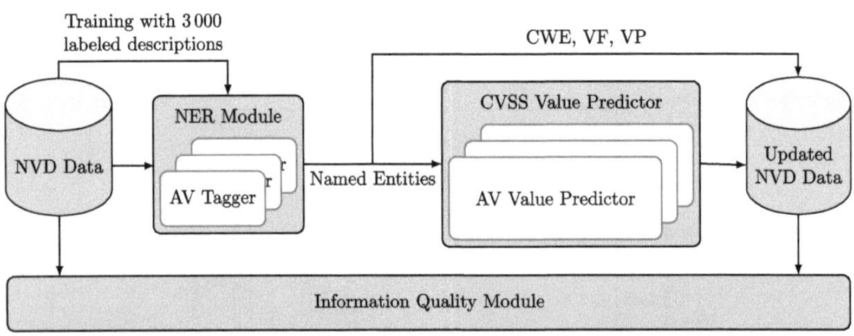

Fig. 11.3 OVANA architecture pipeline

11.3.3.1 Information Quality

Measuring the information quality (IQ) of the NVD requires adequate indicators. Many IQ frameworks are available in the literature [37, 97, 344]. They define different IQ indicators for specific contexts, such as social media [37] or online marketing [344], and categorize them into sub-categories, *e.g.*, subject, object, and process criteria [97]. Since we analyze the IQ of the NVD, we have a specific context with specific requirements. We intend to objectively measure the IQ without using information sources other than the NVD. We selected *accuracy* [38, 39, 224, 334, 338] and *completeness* [38] as IQ indicators to enable a comparison of different approaches in future work. Further, we included the indicator *uniqueness* to measure the uniqueness of vulnerability entries. Unique entries give automated tools a better distinguishability. For quantification, we only use structured information that carries information about the described vulnerability, *e.g.*, the CPE information, while we ignore artificially introduced identifiers such as the CVE-ID. We denote the accuracy, completeness, and uniqueness scores with *acc*, *comp*, and *uniq*, respectively. To get the overall IQ, we simply add *acc*, *comp*, and *uniq* together.

Accuracy Information accuracy has been extensively measured in previous work [38, 39, 334]. However, previous work either ignores the free-form vulnerability description or focuses on extracting specific information, like the software name and software version. Accuracy can be determined by comparing information with other sources, in our case the original with the predicted information. We restrict the accuracy score to a comparison of the original and predicted CVSS score, since comparing different CWE classes is not useful. Eq. (11.2) returns the accuracy of the entire database, where $cvss\colon V \rightarrow [0, 10]$ and $predicted\_cvss\colon V \rightarrow [0, 10]$ indicate the current and predicted CVSSv3 scores of a vulnerability, respectively.

$$acc\colon D \rightarrow \mathbb{R}_{\geq 0} \tag{11.1}$$

$$acc(D) = \frac{\sum_{v \in D} |cvss(v) - predicted\_cvss(v)|}{|D|} \tag{11.2}$$

It basically calculates the averaged absolute Δ over all original and predicted CVSS scores.

Completeness The completeness of information can be determined by identifying missing information in a vulnerability entry. As an example, older entries ($\leq$ 2015) sometimes lack CVSSv3 information. The completeness of a single vulnerability is given by $comp : V \rightarrow [0, |fields|]$, which indicates the number of fields with a value.

Eq. (11.4) shows the completeness formula for the whole range of vulnerabilities in the dataset D. The lower the value of *comp*, the better.

$$comp\colon V \rightarrow [0, |fields|] \subseteq \mathbb{R} \tag{11.3}$$

$$comp(D) = \frac{\sum_{v \in D} c(v)}{|D|} \tag{11.4}$$

Uniqueness The uniqueness of vulnerability entries can be identified by comparing entries with each other, *i.e.*, a vulnerability $v \in D$ is unique *iff* $\forall v' \in D \setminus \{v\} : fields_v \neq fields_{v'}$, where D is the set of all vulnerabilities. Since this information alone misses improvements if large clusters are split into smaller ones, we also incorporate cluster sizes in Eq. (11.6), where $C \subset \mathcal{P}(D)$ is the set of clusters. The smaller the clusters, the smaller the value, indicating a better IQ, with a minimum of 0 if all vulnerabilities are unique.

$$uniq\colon C \rightarrow \mathbb{R}_{\geq 0} \tag{11.5}$$

$$uniq(C) = \frac{\sum_{c \in C} \left(|c|^2\right)}{|C|} - 1 \tag{11.6}$$

All formulas combined give the IQ of the vulnerability dataset D, where the smallest possible and best IQ value is 0. This answers our first research question.

11.3.3.2 Named-Entity Recognition

We want to identify named entities in the free-form descriptions, *e.g.*, words, which are related to a specific CVSS parameter, using NER [333, 345].

The labels in this paper are listed in Table 11.1. The former two are equivalent to the labels of Dong et al. [39], but are not in scope of this paper. They may be used in future work to correct possible inconsistent CPE information. The following eight labels are based on the CVSSv3 base metric, which are used to analyze the accuracy of the CVSS and complete incomplete vulnerability entries (*cf.* Problem 2 in Sect. 11.3.2). The label *weakness* was identified to potentially increase the accuracy of the CWE field (*cf.* Problem 1 in Sect. 11.3.2), while *VF* and *VP* are new fields that increase the *uniq* of entries (*cf.* Problem 3 in Sect. 11.3.2).

11.3.3.3 CVSS Value Predictors

The CVSS score approximates the severity of a vulnerability[8]. It consists of different metrics: the base, temporal, and environmental metric, and of different parameters within these metrics. The NVD covers only the base metric. The other metrics depend on either temporal information that may change over time (temporal metric), or user specific information, like IT infrastructure information (environmental metric). The NVD offers a CVSS base vector, which holds the information used to calculate the severity in a structured way. This vector however, is not always available, yielding the highest severity, or contains data that is inconsistent with the vulnerability description. We try to (i) calculate a CVSSv3 vector for entries, which provide none; and (ii) detect inconsistencies between existing CVSSv3 vectors and the vulnerability description.

For this task, we leverage the NER module to detect named entities in the description associated with the CVSS vector parameter (cf. Table 11.1). Further, we use a neural network to predict the value of a parameter p based on the words related with that parameter. Here, we train one model for each parameter (cf. Table 11.1), resulting in eight different models. In this way, we are able to predict the CVSS parameter values for NVD entries with an outdated or non-existent base vector and detect inconsistencies with the description of the original vector. To measure these inconsistencies, we use the *acc* of the IQ module, which calculates the difference of the original CVSS score s_o and the predicted CVSS score s_p, *i.e.*, $|s_o - s_p|$.

In contrast to the machine learning model of Anwar et al. [38], our pipeline is capable of predicting the whole vector, updating old values, correcting incorrect values, and suggesting a vector for newly added entries. In contrast to the work of Elbaz et al. [224], we use deep learning algorithms with state-of-the-art embeddings [47] instead of a linear model with a bag of words approach. The advantage of their model is that no additional labeling of the data is required, which can be beneficial in terms of generalization and variance minimization in machine learning. On the other hand, a more powerful algorithm can reduce the machine learning bias error, as in our case. Despite our more sophisticated model, we can also guarantee some form of explicability since the first part of the pipeline identifies named entities that lead to the prediction.

[8] https://www.first.org/cvss/specification-document

11.4 Evaluation

This section presents the implementation of the ML models, the evaluation of those, and the results of the IQ analysis.

11.4.1 Architecture and Implementation

The NER module consists of one tagger module for each tag (cf. Table 11.1). This way, the NER module is able to predict multiple tags for a given word. On top of each CVSS tag (all CVSS tags in Table 11.1), we build an additional classifier that predicts the values of the tagged words.

We base our NER tagger and CVSS value predictor module on deep neural networks. Since neural networks can only work with numbers and vectors and not with natural language, we need to encode words. Context-sensitive word embeddings are the state-of-the-art in this field. We use BERT embeddings [47] and the Flair NLP framework [333], which are stacked together by concatenation. For the NER module, we feed those embeddings into a bidirectional long short-term memory (LSTM) network with a Conditional Random Field (CRF) layer as in the work of Huang et al. [346]. On top of this, the output nodes are fully connected to the LSTM hidden state.

To train a predictor model for a CVSS parameter p, we input all word embeddings, labeled with its tag in the vulnerability description and the corresponding CVSS value given in the NVD entry, and feed them into a gated recurrent unit (GRU) [347]. If a description does not contain information about a CVSS parameter, that description is skipped in the training process. In the end, a fully connected layer connects the hidden state to the output nodes.

11.4.2 Hyperparameters, Dataset Selection, and Evaluation Metrics

The models of the taggers and classifiers are very similar, so we use similar hyperparameters. Each tagger and classifier is trained using stochastic gradient descent (SGD) with a mini batch size of 32 and cross entropy loss. We use a learning rate of 0.1 with a anneal factor of 0.5 and limit learning to a maximum of 20 epochs. The embeddings are fine-tuned within each task. The language model of the flair embeddings has a dropout rate of 0.25, while the transformer architecture of BERT uses a dropout rate of 0.1. From the recurrent neural network (RNN) layer (LSTM or GRU) to the final layer dropout with a rate of 0.5 is applied. On the development set, we tried several other hyperparameter settings. *E.g.*, we noticed that the higher the number of epochs, the better the results. We monitored the micro F1 score on the development set for 5, 15 and 20 epochs. As a result, the availability impact classifier, for example, had a score of 0.6704, 0.6592, and 0.6536. For the learning rate, we performed the learning rate range test [348] for several taggers and classifiers, of which we identified 0.1 as a reasonable value. We also tried the ADAM [210] optimizer instead of SGD, but this produced worse results for the taggers. A larger batch size was not possible due to computational restrictions. Encountering a specific type of errors (explained in the next subsection), we also experimented with the tagging scheme. We tried the `iob2` tagging scheme [349], which, however, did not lead to significantly better results.

Furthermore, we noticed a high imbalance in the data for the training tasks. To address this issue, we implemented two versions of oversampling (OS). One is used in the NER module to oversample the training descriptions that contain the specific tag, while the other is used for the CVSS value predictors to oversample the phrases that contain the specific CVSS value. The OS functions are parameterised by $n \in \mathbb{N}_{\geq 1}$, denoting the repetition factor for the specific data instances. The different OS factors are given within the results of the evaluations.

For both evaluation tasks, we split the annotated data 80%, 10%, and 10% into a training-, development-, and testing-set. Unlike many other NER-tasks which split the data sentence-wise, one entry of the data set is seen as a whole NVD description. This way, the neural network can integrate the context of the whole entry without being limited by sentence boundaries. While we kept the same splits for the NER task, we performed a new split for each value prediction task so that the multiple classes of each tag are evenly distributed.

Each evaluation result is presented with the metrics accuracy, recall, F1 score, and the weighted average.

11.4.3 NER Module

An extract of the tagger results from the NER module can be found in Table 11.2 (for the full overview, see Appendix Table A.2). It becomes apparent that the weighted average of all taggers is high. However, the weighted average mostly represents the O tag. While it is important for the performance of a whole tagger, we are also interested in the performance for individual tags. From this perspective, we can observe that the tagger results are mixed. For some cases, *e.g.*, in predicting software name (SN), software version (SV), and user interaction (UI), the tagger results are very good, with a F1 measure of 0.8584, 0.9047, and 0.9359, respectively. However, there are cases where the tagger performs worse, *e.g.*, when tagging scope (S) or attack complexity (AC) with F1 scores of 0.5455 and 0.6136, respectively.

Table 11.2 NER tagger module evaluation

	Tag	Accuracy	Recall	F1
Software name	SN	0.8975	0.8226	0.8584
OS: n = 0	O	0.9792	0.9889	0.9841
Weighted Average		0.9706	0.9713	0.9708
Software version	SV	0.8516	0.8965	0.8735
OS: n = 0	O	0.9892	0.9838	0.9865
Weighted Average		0.9762	0.9756	0.9758
User interaction	UI	1.0000	0.8795	0.9359
OS: n = 0	O	0.9994	1.0000	0.9997
Weighted Average		0.9994	0.9994	0.9994
Scope	S	0.3750	1.0000	0.5455
OS: n = 50	O	1.0000	0.9994	0.9997
Weighted Average		0.9998	0.9994	0.9995
Attack complexity	AC	0.6575	0.6005	0.6277
OS: n = 2	O	0.9725	0.9784	0.9754
Weighted Average		0.9521	0.9539	0.9529

As we inspected these errors, we noticed that some taggers are not able to detect the exact boundaries of tags. For example, see the SV tagger prediction in the description of CVE-1999-0131[9]: *"[..] denial of service in Sendmail 8.7.5 <SV> and <SV> earlier <SV> through <SV> GECOS field [..]"*. While the tagger correctly tags *"8.7.5"*, *"and"*, and *"earlier"* as software version, *"through"* is incorrectly recognized as such. For this reason, we also tested iob2 tagging [349], where tags are supplemented with *beginning (B)* and *inside (I)* tags, which support the CRF layer due to the potentially decreased search space. Unfortunately, this did not lead to better results. However, it should be mentioned that a recall and accuracy of more than 60% is still far from a poorly learned or random behavior, since extracting the correct words in a large set of words is a difficult task. This is the reason

[9] https://nvd.nist.gov/vuln/detail/CVE-1999-0131

why we introduced OS, because otherwise the classifier would sometimes only predict O-tags, since the interesting tags rarely occur. Moreover, the results suggest that the tags that occur more frequently seem to be better predicted by the taggers. This implicates that more annotated training data leads to better performances of the taggers. The quality measurements are also influenced by incorrect tagger behavior. In contrast to the recognition of random words, as described in the error analysis, errors such as tagging too many words are not particularly serious. In the end, we only expect the NER module to find the correct tags within the descriptions. The exact limits at which the tags appear in the text are helpful, *e.g.*, if we want to make the classifications more explainable.

11.4.4 CVSS Value Predictors

Each CVSS value predictor is a neural network suitable for multi-class prediction of the values of one extracted CVSS tag. This way, we have, *e.g.*, a specific classifier suitable for the Attack Vector (AV) tag and able to differentiate between the associated values "network", "adjacent", "local", and "physical".

Similar to the results of the taggers, the classifiers performances are also scattered; an extract of the results is given in Table 11.3 (for the full overview, see Appendix Table A.2). With the help of the OS technique, the results for the confidentiality impact (CI) and UI classifiers are good, with a weighted F1 score of 0.7105 and 0.8863, respectively. However, there are also classifier that perform worse, despite OS. *E.g.*, it seems rather difficult for the privileges required (PR) or integrity impact (II) classifier to predict the correct value (weighted F1 scores of 0.6386 and 0.6411, respectively).

In contrast to the NER tagger results, no obvious error patterns were found. Nevertheless, it can be assumed that the classifier performance is also increased when more training data is available, as the classifiers can then generalize better. *E.g.*, the classifier for II does not produce good results for the class "None". This class is severely underrepresented, with even OS not yielding satisfying results. We assume that a simple OS is not sufficient, since the instances of this class are highly diverse. This problem might require a more sophisticated data augmentation technique or more training.

Table 11.3 CVSS value classifier evaluation.

	Value	Accuracy	Recall	F1
Confidentiality impact	L	0.7059	0.5373	0.6102
OS: n = 0	N	0.6400	0.6400	0.6400
	H	0.7395	0.8544	0.7928
Weighted Average		0.7152	0.7179	0.7105
User interaction	R	0.8889	1.0000	0.9412
OS: n = 2 (N)	N	1.0000	0.5000	0.6667
Weighted Average		0.9111	0.9000	0.8863
Privileges required	L	0.5714	0.6667	0.6154
OS: n = 2 (H)	N	1.0000	0.5556	0.7143
	H	0.5000	0.6667	0.5714
Weighted Average		0.6984	0.6296	0.6386
Integrity impact	L	0.8056	0.4603	0.5859
OS: n = 3 (N)	N	0.1429	0.1250	0.1333
	H	0.6396	0.8554	0.7320
Weighted Average		0.6817	0.6558	0.6411

11.4.5 Information Quality Analysis

With the IQ module, we analyze the dataset using all structured information available, including ones we added (CPE, CWE, CVSS, VF, and VP). Further, we identify named entities in descriptions using the NER module, which are then handed over to the CVSS predictor module. For the CVSS scores, which are mainly used for *acc*, we went for a conservative approach: If we found no information about a CVSS parameter and therefore cannot predict a value, we fall back on the original NVD dataset. If no information is available here either, we used the worst-case assumption implemented by the NVD[10]. We aim at rectifying the information in the NVD based on the free-form description. Hence, we assume the predicted information to be more aligned with the real-world severity [341], which is the reason the *acc* is added to the IQ score of the original dataset. The values for VF and VP are implemented directly into the updated dataset in structured form, since this information is not available beforehand.

[10] https://nvd.nist.gov/vuln-metrics/cvss

Figure 11.4a shows the overall IQ with the different indicators in a stacked form over the years 1999 to 2019 (which is further broken down in Fig. 11.4b and Fig. 11.4c). One can see that the IQ improved drastically after implementing the additional information and resolving clusters, indicated by a lower IQ score. We lowered the average IQ score from ≈ 5.81 to ≈ 2.86, which is an average decrease by 51.2%. Figure 11.4d depicts the differences of the IQ on an annual basis.

Completeness In Fig. 11.4b the original *comp* shows an average of ≈ 2.78. It can be seen to remain at a mean of ≈ 2.97 until (and including) 2015, after which it drops to a mean of ≈ 2.01. The reason for this trend is the introduction of the CVSSv3 in 2015 and its adoption by the NVD at the end of 2015. All NVD entries after 2015-12-20 are published with a CVSSv3 score, which is reflected in the data. Further, the dataset lacks structured information about affected paths and functions, resulting in a base *comp* of 2. In the updated dataset, *comp* drops to an average of ≈ 1.73, *i.e.*, showing a decrease by 36.31% (*cf.* Fig. 11.4d). The improvement can

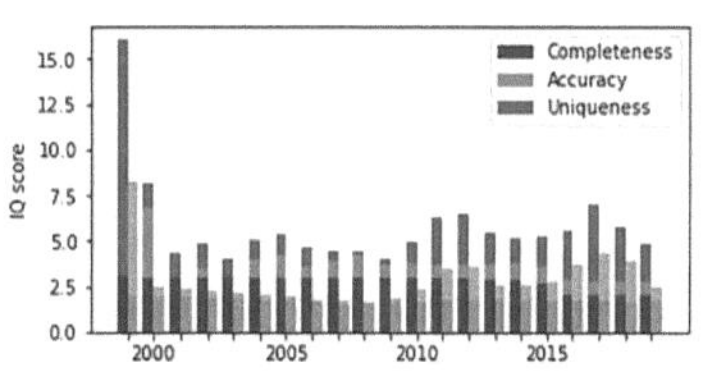

(a) Stacked IQ compared: For each year, the first bar represents the IQ for the original NVD dataset, while the second one represents the IQ of the updated NVD dataset. Note: the lower the score, the better the IQ (*cf.* Section 11.3.3.1). The depicted indicators are $comp$, acc, and $uniq$.

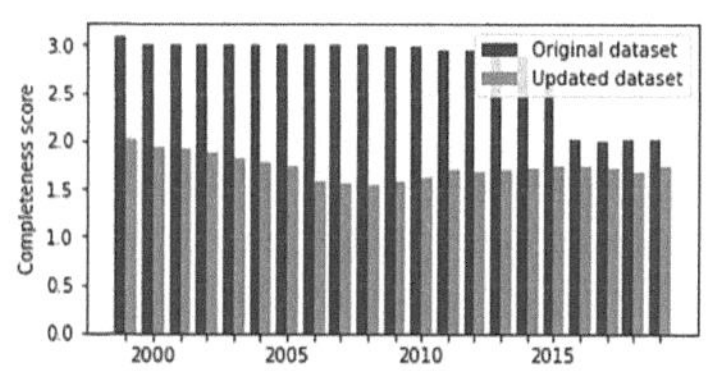

(b) Completeness compared per year. Note: the lower the score, the better the IQ (*cf.* Section 11.3.3.1).

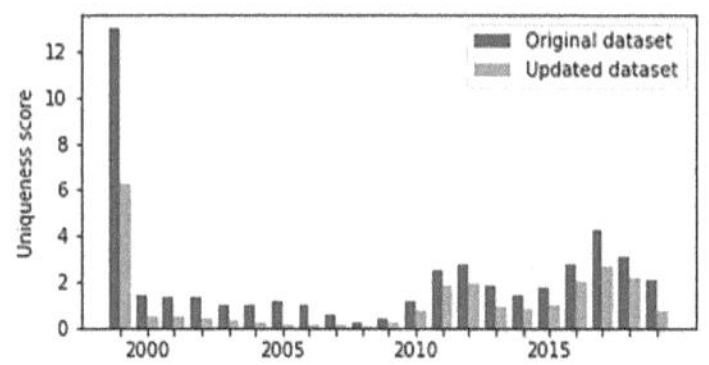

(c) Uniqueness compared per year. Note: the lower the score, the better the IQ (*cf.* Section 11.3.3.1).

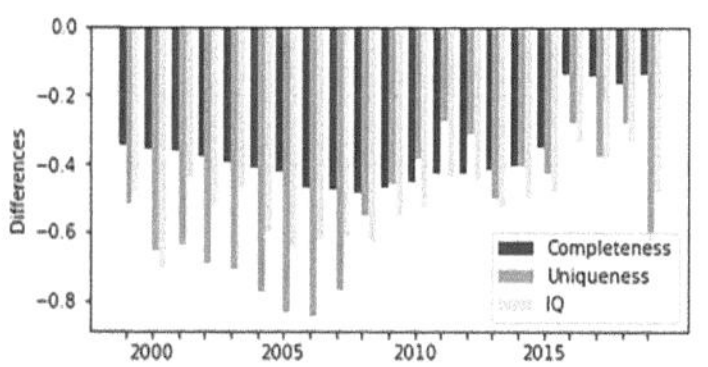

(d) Relative deviations per year of $comp$, $uniq$, and IQ.

Fig. 11.4 Overview of the different information quality metrics and the combined result

be explained by our conservative approach. This way, each vulnerability is assigned a score, eliminating all missing CVSSv3 scores. Additionally, more vulnerability descriptions include information about VFs or VPs, reflected by the slow decline from 1999 until 2008. The remaining missing fields are VF and VP information and the CPE entries, which, however, are beyond the scope of this paper.

Uniqueness The *uniq* in Fig. 11.4c shows a peak in 1999 of the original dataset at $\approx$ 12.94, but averages out thereafter with a value of $\approx$ 1.63 excluding 1999 (at $\approx$ 2.17 including 1999). One can see two peaks in 2012 and 2017 of the original dataset with $\approx$ 2.76 and $\approx$ 4.26, respectively. The peak in 1999 arises from a widespread lack of structured information (`cpe`, `cwe`, and `cvss`) and forms a cluster of 117 entries. In 2012, there was one specific cluster with 45 vulnerability entries for `Apple iTunes` with no `cwe` and CVSS specification. These 45 entries are in a range of 123 ids, indicating some sort of batch-add, *i.e.*, all of which were added within a day or two. Similarly, in 2017 there was a cluster for `tcpdump`[11] which contained a very broad CPE specification, the same CWE class and the same CVSS vector for all entries. Due to their consecutive IDs, these vulnerabilities are also probably batch-added. Some of the batch-added entries even have the same description (*e.g.*, the largest cluster in 2011), which is problematic for the NER module. With such descriptions, the NER module is unable to identify distinctive information. The updated dataset shows an average *uniq* of 1.12%, *i.e.*, a decrease by 53.85% (*cf.* Fig. 11.4d). Relatively speaking, the original and updated *uniq* look quite similar. This results from the aforementioned problem of the NER module with equal descriptions, which is common in large clusters.

11.5 Discussion

This section presents the contributions of the research conducted in this paper and discusses its limitations.

11.5.1 Contributions

The contributions of this paper are discussed in the following section. We categorized them in theoretical and practical contributions.

[11] https://www.tcpdump.org/

The first theoretical contribution is the **definition of an objective IQ measurement for vulnerability databases**, based on insights in different IQ frameworks [37, 97, 344] (**T1**). To our best knowledge, no published study introduced such a measurement. Previous work used self-contained measurements bound to their use case [38, 39, 224]. Our approach allows for a comparison of different approaches, independent of the approach itself. The second theoretical contribution is the **proposed machine learning pipeline (T2)**. It consists of a NER tagger (*cf.* Sect. 11.3) that is able to identify relevant CVSS information, and CVSS predictors which predict the value of a CVSS parameter based on the identified named entities. While Elbaz et al. [224] use linear models with a bag-of-words approach and need no labeling effort, we use deep learning models with embeddings and manually labeled data. Our pipeline overcomes the limitation of the linear model, but is restricted in the data variety since the training data needs to be manually labeled. Both approaches try to find a way to make the prediction reasonably explainable, which is an important topic in machine learning research and especially important for the CVSS score prediction, since the manual assignment of it lacks transparency.

Our proposed tool is able to **identify additional information in the free-form descriptions and decreases the cluster sizes** within the NVD (**P1**). This increases the uniqueness of each vulnerability and speeds up the search for vulnerabilities in source code. Moreover, it does allow faster reproducibility of vulnerabilities and consequently faster patches. Previous work [39] used similar schemes to identify named entities, but focused on singular pieces of information, *e.g.*, software name or software version, while our approach captures further information (VP, VF, and CVSS information). One important thing to note here is that more information does not necessarily make everything more secure. This has been discussed by You et al. [169] as they collected semantic information about vulnerabilities on references to automatically generate exploits. Hence, we only include information which was already available in the NVD, just not in a structured form. Our tool is potentially able to **guide security experts and practitioners during CVSS assessment (P2)**. Immediately after a vulnerability is publicly disclosed, usually no information is available besides a free-form description and references, if any [224]. Our proposed tool is able to assist security experts in finding an appropriate CVSS score and help practitioners assess the severity of a vulnerability when no score is available. Further, it provides an unbiased suggestion, *i.e.*, individual parameters are handled separately. The machine learning pipeline consists of a NER tagger, which, among other things, is able to identify the specific CVSS vector fields, and a CVSS value predictor, that classifies the found fields into the appropriate CVSS values. In addition, the pipeline enables a form of interpretability of the CVSS prediction which is helpful when a score already exists but it is not clear on which basis the score was assigned. This leads to the next practical implication. As mentioned earlier, the CVSS has been highly criticized for its non-transparent scoring system (*cf.* Sect. 11.2). Our proposal attempts to **increase the transparency of the scoring (P3)**. The tool is split into two layers of machine learning models. Hence, its interim results, *i.e.*, the key words that led to the CVSS prediction, can be presented to the user by highlighting its associated NER tag, CVSS tag, and predicted CVSS value. This can assist the security expert in CVSS rating, by offering some degree of explainability. Enriched with the descriptions of the respective tag and value, the information is also essential for practitioners seeking further information about severity classification.

11.5.2 Limitations and Future Work

Despite the promising results, our paper has some limitations that could be addressed in future work in this area.

Information Quality This paper evaluates the IQ based on the indicators of accuracy, completeness, and uniqueness. The selection is based on research conducted in previous work in this field. In order to gain a more universal insight into IQ, additional indicators need to be included into the calculation. Further, we analyzed the IQ without the inclusion of results of analysis done in previous work, if they are not already included in the NVD [38, 39]. Future work could increase the number of indicators used for the IQ assessment and could evaluate the IQ improvements of previous inconsistency analyses. Furthermore, future work could look deeper into the topic of automatic CPEs generation based on vulnerability reports, as there are still problems as indicated in Sect. 11.3.2, and adjust this method to additional vulnerability databases.

Machine Learning Our results show a high result fluctuation of the NER and CVSS modules. Further, there is a correlation between the quality of the results and the amount of training data available. Since we use manually labeled data, future work could create a crowdsourcing task to gain more training data. Another approach to artificially increase the size of the training data set is to incorporate sophisticated data augmentation techniques **[bayer-data-2021**, [350] to increase the size of this training data set. If only the CVSS score is of interest, it would be possible to learn a regression model that does not require manual labeling because the score can be retrieved from the NVD.

11.6 Conclusion

Securing IT infrastructures is becoming more difficult every year. Vulnerability databases are one of many information sources for IT security experts to publish and learn about new threats. Previous work has shown inconsistencies in vulnerability databases and mixed results in terms of reproducibility, which calls for automated tools to obtain more information about published vulnerabilities and improve the quality of the information available. We answered our first research question, how to assess the IQ of vulnerability databases, by defining an IQ metric based on the indicators of accuracy, completeness, and uniqueness. A preliminary analysis of the NVD revealed large clusters of vulnerabilities sharing the same structured information, but with distinguishable descriptions, containing further information about the vulnerabilities. Hence, we answered our second research question, on which information improves the IQ of the NVD by using a tool called OVANA which analyzes the IQ of the NVD using state-of-the-art NLP and ML. It outlines important words and terms in the free-form descriptions of the vulnerability entries and implements them in the vulnerability dataset. With this technique, we were able to improve the IQ of the NVD by 51.23%. Additionally, our tool is able to guide IT security experts in selecting the correct values for CVSS parameters.

12 Common Vulnerability Scoring System Prediction Based on Open Source Intelligence Information Sources

Abstract

The number of newly published vulnerabilities is constantly increasing. Until now, the information available when a new vulnerability is published is manually assessed by experts using a Common Vulnerability Scoring System (CVSS) vector and score. This assessment is time consuming and requires expertise. Various works already try to predict CVSS vectors or scores using machine learning based on the textual descriptions of the vulnerability to enable faster assessment. However, for this purpose, previous works only use the texts available in databases such as National Vulnerability Database. With this work, the publicly available web pages referenced in the National Vulnerability Database are analyzed and made available as sources of texts through web scraping. A Deep Learning based method for predicting the CVSS vector is implemented and evaluated. The present work provides a classification of the National Vulnerability Database's reference texts based on the suitability and crawlability of their texts. While we identified the overall influence of the additional texts is negligible, we outperformed the state-of-the-art with our Deep Learning prediction models.

Original Publication ***Kuehn, P.***, *Relke, D. N., Reuter, C., "Common vulnerability scoring system prediction based on open source intelligence information sources,"* Computers & Security, *2023.* https://doi.org/10.1016/j.cose.2023.103286.

| IF 4.8[1]

[1] Updated 5. Februar 2025.

Minor changes regarding language, style, and formatting have been made in this chapter to provide a consistent experience.

P. Kühn, *Proactive Cyber Threat Intelligence*, Technology, Peace and Security | Technologie, Frieden und Sicherheit,
https://doi.org/10.1007/978-3-658-49241-0_12

12.1 Introduction

IT systems are now ubiquitous and fundamental to society, businesses, and individuals. Failures and disruptions can have catastrophic consequences for those affected. In 2017, for example, two waves of ransomware attacks occurred, each resulting in major outages to businesses and infrastructure [124, 351]. The vulnerability that enabled these attacks had been known and fixed a month before the first attack. In other attacks, such as the one on Microsoft Exchange Server in early 2021, only a few days passed between the discovery of the vulnerability and the start of attacks [352].

It is therefore important for researchers or system administrators to learn about vulnerabilities as early as possible, analyze them and initiate countermeasures. Various publicly accessible databases, such as the National Vulnerability Database (NVD)[2] and the Common Vulnerabilties and Exposures (CVE)[3] collect, structure and prepare the published vulnerabilities for this purpose. However, relevant information can also be found on many other platforms, such as social media (especially Twitter), blogs, news portals, and company websites.

The Common Vulnerability Scoring System (CVSS) is used to categorize different aspects of vulnerabilities. The result of this categorization is a vector whose elements are a machine-readable representation of the vulnerability's properties[4]. Based on the components of the CVSS vector a numerical vulnerability score (CVSS severity score) is calculated. The vulnerability assessment is usually performed by IT security experts based on the available open-source intelligence (OSINT) information. OSINT refers to the structured collection and analysis of information that is freely available to the public.

There is a certain period of time when the information about a new vulnerability is published, but the assessment made by experts is not yet available [224, 343]. Due to the large mass of published vulnerabilities, it is difficult for researchers or, *e.g.*, responsible persons in companies to assess each new vulnerability themselves. They are therefore dependent on the assessments of experts. Accordingly, the longer it takes for the assessment to become available, the longer it takes for countermeasures to be taken to mitigate the vulnerability. During this period, the vulnerable systems are vulnerable to attack without the responsible parties knowing about it. It is therefore important that the assessment is available as soon as possible.

Various works [224, 353, 354] try to perform this assessment automatically based on the textual information available about a vulnerability using machine learning

---

[2] https://nvd.nist.gov/

[3] https://cve.mitre.org/

[4] https://www.first.org/cvss/specification-document

(ML). This would allow for a much faster assessment. The vulnerability could already be assessed in an automated way when it is published and the time window in which no at least preliminary assessment is available is kept small. It would also allow experts to prioritize and make recommendations for the assessment.

Previous work largely uses only the short descriptions of vulnerabilities from NVD and CVE with some exceptions [276, 355]. Han et al. [353], for instance, present a system for classifying vulnerabilities into different severity levels based on CVSS. From Khazaei et al. [356] comes a work on predicting the numerical CVSS severity score. In addition, there are methods that automatically predict the entire CVSS vector [224]. Another work by Kuehn et al. [5] describes a system that uses Deep Learning to predict the CVSS vector. However, the system requires labels created by experts to train, which significantly increases the required effort for larger datasets. Further, Deep Learning (DL) profits from large training datasets to which the reference texts could contribute, which is currently not leverage by related work.

Goal This work aims to use as much textual data as possible to predict the CVSS vector of a vulnerability. This is to achieve the most accurate estimation of the CVSS vector possible. It should be possible to use not only the short description of the vulnerability, but also other types of texts, such as Twitter posts and news articles for prediction in case of a new vulnerability. Possible sources of textual information about vulnerabilities should be found and categorized. We aim to answer the following research questions:

RQ1 **Where can relevant textual information on vulnerabilities be found outside vulnerability databases?**

RQ2 **To which degree are public data sources beyond vulnerability databases suitable for predicting the CVSS vector?**

This will clarify whether there are typical sources that regularly report on current vulnerabilities and whether these are suitable as a basis for building a dataset for training a ML system.

Here, a first impression shall be gained by a rough manual search and then the sources referenced in the databases shall be analyzed automatically with regard to the type and scope of the references (*e.g.*, blog posts, patchnotes, GitHub issues). With the help of the texts, a ML model for predicting the CVSS vector is to be trained. The data must be filtered and cleaned for this purpose. The ML model shall use Deep Learning and use state-of-the-art models as a basis. The model is evaluated and compared to previous work.

Contributions The contribution to current research is an analysis of the references contained in the databases. This will categorize the references in terms of certain characteristics and suitable for ML models and can serve as a starting point for further work on the use of the references **C1**. A method that collects and processes the text contained on the referenced web pages will be presented. In addition, a system is implemented and evaluated that, unlike previous work, such as Elbaz et al. [224] and Kuehn et al. [5], uses more extensive text from the references in addition to descriptions of vulnerabilities from the databases **C2**. This method for predicting CVSS vectors surpasses the current state-of-the-art. Further, we present an extensive explainability analysis of our trained models as part of our evaluation **C3**.

Outline The state-of-the-art in research is considered in Sect. 12.2, followed by a preliminary analysis of the references included in NVD (*cf.* Sect. 12.3). Requirements for references and the texts contained in them are defined and consequently the individual references are evaluated, resulting in a selection of references. Sect. 12.4 explains the procedure for collecting the texts from the references and a system for retrieving, processing, and storing the texts is presented. Sect. 12.5 evaluates the ML system, while Sect. 12.6 discusses and compares the results with other work. Finally, a conclusion is drawn in Sect. 12.7.

12.2 Related Work

This section gives an overview over the state-of-the-art in research. We focus literature dealing with the prediction of CVSS vectors, scores, or levels. In addition, work that uses sources other than NVD in this context is considered. Automated assessment should provide a time advantage over the assessment by human experts. In this regard, different papers come to different conclusions regarding the duration of the assessment, and the exact methodology is not always clear. Elbaz et al. [224] state for the observed period from 2007 to 2019 that 90% of vulnerabilities were assessed within just under 30 days, with a median of only one day, while Chen et al. [355] indicate an average of 132 days between publication and assessment for an observed period of 23 months in 2018 and 2019.

NVD, CVSS, Information Sources Johnson et al. [332] perform a statistical analysis of CVSS vectors in different databases containing vulnerabilities. In doing so, they show that despite different sources, the CVSS vector is always comparable and, consequently, seem to be robust. They state the NVD is the most robust information source for CVSS information. On the other hand, Dong et al. [39] show that information in the NVD itself is sometimes inconsistent and propose a system that relies on external sources to find, for example, missing versions of the software in question in the NVD. Accordingly, Kuehn et al. [5] present an information quality metric for vulnerability databases and improve several drawbacks in the NVD. In addition to vulnerability databases, other sources of information are used in vulnerability management. Sabottke et al. [125] use Twitter to predict whether a vulnerability will actually be exploited. Almukaynizi et al. [276] go a step further and use other data sources, such as ExploitDB[5] and Zero Day Initiative[6]. However, no text is used, but the simple existence of an article about a vulnerability is used as a feature for the ML model.

CVSS Prediction Numerous works deal with the prediction of CVSS vector, scores, or levels starting from text. As one of the first works, Yamamoto et al. [357] use sLDA [358] to predict the CVSS vector based on the descriptions. For predicting the score, Khazaei et al. [356] use Support Vector Machines (SVMs), random forests [359], and fuzzy logic. Spanos et al. [360] predict the CVSS vector using random forests and boosting [361]. DL is first used in this context by Han et al. [353]. By using a Convolutional Neural Network (CNN), no feature engineering is required. However, in doing so, the model only determines the CVSS severity level from the options *Critical*, *High*, *Medium*, and *Low*. Gawron et al. [362] use DL in addition to Naive Bayes, but here the result is a CVSS vector. Twitter serves as the data source for Chen et al. [176]. The ML model is based on long short-term memory (LSTM) [363] and predicts CVSS score. Sahin and Tosun [364] also improve on the Han et al. [353] approach by using a LSTM. Gong et al. [365] show a multitask learning method that sets up multiple classifiers on a single neural network (NN), making it more efficient. Liu et al. [366] use the Chinese equivalent, the China National Vulnerability Database of Information Security (CNNVD), as the data source rather than the NVD. Jiang and Atif [367] take scores not only from the NVD but also from other sources as a basis for their prediction of the score. The work of Elbaz et al. [224] focuses on a particularly tractable classification of the CVSS vector. Therefore, they do not use dimension reduction techniques. Kuehn et al. [5] use DL to predict the CVSS vector, based on the NVD's descriptions, with the goal to aid security experts in their final decision. The most recent approach proposed Shahid and Debar [354], which uses a separate classifier based on a Bidirectional Encoder Representations from Transformers (BERT) model [47] to determine the CVSS vector for each component of the vector. Several proposals rely solely on the textual data from the NVD. Some use text from Twitter or simple binary features,

[5] https://www.exploit-db.com/

[6] https://www.zerodayinitiative.com/

such as the existence of an article about a particular vulnerability. Other vulnerability context tasks also use few different data sources. Yitagesu et al. [368] also use Twitter as a source for a model for Part-of-Speech (POS) tagging. Liao et al. [33] propose a system which draws on several sources to filter Indicator of Compromise (IoC) from natural text.

Research Gap OSINT is widely used in IT security [33, 125, 175, 249]. Various works exist on the prediction of CVSS vectors based on descriptions. However, as research shows, few OSINT vulnerability sources are used [86], especially in the context of CVSS score, level, or vector prediction, and if they are, very simple features from other sources are used [276]. Furthermore, there is no systematic analysis of the suitability of NVD references for CVSS vector prediction approaches.

12.3 Preliminary Analysis

The authors performed an exploratory analysis of the available data, *i.e.*, vulnerability descriptions and outgoing references from the NVD, to identify data suitability criteria and requirements for web scraping. Suitable in the sense of the present work are texts that describe a vulnerability and can be directly assigned to a vulnerability via the CVE identification number. Some special factors have to be considered:

- Each text shall be uniquely assignable to one and only one vulnerability via the CVE identification number. Without this criterion a text could be used as a training example for two different permutations of one of the components of the CVSS vector. This makes it difficult for the ML algorithm to identify the relevant properties of the vulnerability. Also, since the vulnerabilities covered in a text may be very different, it does not make sense to use the same text for multiple vulnerabilities. It is even possible that only one vulnerability is really described, although several with different target vectors are mentioned.
- The texts should not contain the target variable, *i.e.*, the CVSS vector. Otherwise, the ML model could predict the target parameter based on the variable present in the input, without any actual meaningful learning effect.
- There should be as little noise as possible. This ensures a high quality of the prediction. As stated in Sect. 12.2, the data otherwise contain patterns that could negatively affect the ML model.

Our secondary goal with this exploratory analysis is to identify *where to find usable data*, *assess the data quality* and *how it can be used*. Those questions correlate with our research questions (*cf.* Sect. 2).

12.3.1 Descriptions in the NVD

The first and most important starting point for finding texts about vulnerabilities is the NVD. We consider NVD entries from 2016 to 2021, based on the introduction of the current CVSS standard version 3. Entries without CVSS version 3 information are excluded. This is the case for vulnerabilities in 2016, when CVSSv3 was still in the process of wide adoption, and in 2021, where the CVSSv3 vector was not yet available at the time the entries were retrieved. In total, we collected 88 979 entries.

Individual entries in the NVD contain a short, expert curated[7] description of the vulnerability. The length of the descriptions for our collected entries ranges between 23 and 3 835 characters, with an average of 310 and a median of 249. Fig. 12.1 shows the distribution of the length of the descriptions. Descriptions longer than 1 000 characters are very rare, with the 95th percentile already at 746 characters. The information content of texts correlates with the pure length of the texts, apart from some exceptions[8]. Likewise, a single, short sentence cannot describe all aspects of the vulnerability. As Fig. 12.1 illustrates, there are lots of vulnerabilities in the NVD with very short descriptions.

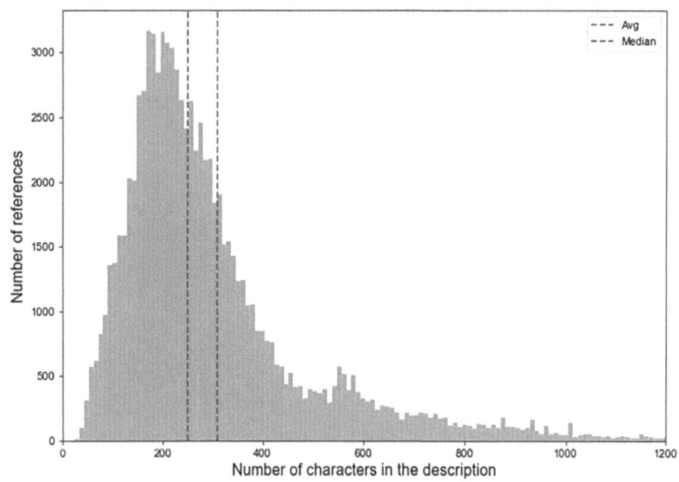

Fig. 12.1 Distribution of National Vulnerability Database description lengths

[7] https://www.cve.org/ResourcesSupport/FAQs#pc_cve_recordscve_record_descriptions_created

[8] Some descriptions list other, non-identical, vulnerabilities, which artificially increases the length of the description without giving further content.

Literature shows that the quality of vulnerability descriptions in the NVD differs [5] and the quality can only be assessed to a limited extent without a deeper analysis. A random sample shows that many descriptions contain less information about the actual vulnerability, but list, *e.g.*, affected products and version numbers. Such information is unrelated to the characteristics of the vulnerability and is therefore of little usefulness to predict the vulnerability severity. Nevertheless, Shahid and Debar [354] show that good results in the prediction of the CVSS vector are possible based only on NVD descriptions. Their method of CVSS score prediction achieves a Mean Squared Error (MSE) of 1.79 and a correctly predicted score in 53% of all cases.

12.3.2 Reference Analysis

Each NVD entry references websites. To identify, which websites are suitable to be crawled we first analyze what kind of references are involved and, based on these insights, build categories for reference domains. Second, we rate these groups based on their *crawlability* and potential text quality.

In the given subset of all entries of the NVD there are a total of 251 485 references. The median number of references per vulnerability is 2. Many vulnerabilities have only a single reference, 95% have 8 or fewer references. There are a few outliers with over 100 references. The distribution of the number of references can be seen in Fig. 12.2.

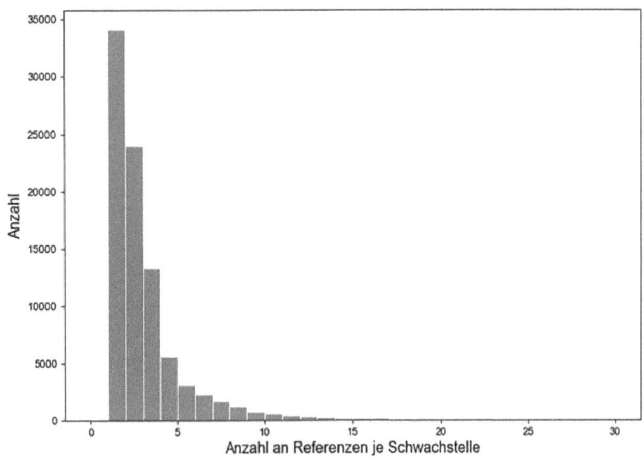

Fig. 12.2 Distribution of number of references per vulnerability in National Vulnerability Database

Over time, the diversity of references increased slightly. From 2016 until 2021 there are 6 013 different referenced domains, of which about 75% are accounted for by the 50 most frequent ones. In 2016, 990 different domains are referenced with 86% of all references coming from the 50 most referenced websites. In 2021, this trend increases to a total of 1 711 domains referenced and the top 50 account for 74% of all references, showing an increase of diversity. We build the 100 most frequently referenced NVD reference domains based on our dataset (*cf.* Table 12.1). These domains account for 83% of all references in NVD.

Table 12.1 shows the 30 most frequently referenced domains, with additional entries to properly represent each group.

Table 12.1 The 30 most referenced domains in NVD, with additional entries to properly represent each identified group. # gives the position in the Top-100, *Num.* refers to how many times the given domain is referenced, *Gr.* gives the assigned group, and *Avail.* depicts, whether unavailable domain is unreachable, redirect to domains unrelated to the vulnerability, or are only reachable after login

#	URL	Num.	Gr.	Avail.
1	github.com	25064	1	✓
2	www.securityfocus.com	20 645	2	✓
3	www.securitytracker.com	10 842	2	✓
4	access.redhat.com	8 627	3/4	✓
5	support.apple.com	8 069	3	✓
6	lists.opensuse.org	7 930	2	✓
7	lists.fedoraproject.org	7 212	2	✓
8	www.oracle.com	7 006	3	✓
9	lists.apache.org	6 294	2	✓
10	www.debian.org	5 614	2/3	✓
11	security.gentoo.org	5 289	4	✓
12	usn.ubuntu.com	5 225	3	✓
13	lists.debian.org	4 921	2	✓
14	portal.msrc.microsoft.com	4 391	3	✓
15	www.openwall.com	4 136	2	✓
16	packetstormsecurity.com	4 068	4	✓
17	source.android.com	3 672	3	✓
18	seclists.org	3 462	2	✓
19	www.exploit-db.com	3 412	5	✓

(Continued)

Table 12.1 (Continued)

#	URL	Num.	Gr.	Avail.
20	tools.cisco.com	3 019	4/5	✓
21	security.netapp.com	2 890	5	✓
22	www.ibm.com	2 807	4	✓
23	exchange.xforce.ibmcloud.com	2 673	4	✓
24	helpx.adobe.com	2 643	3	✓
25	www.zerodayinitiative.com	2 547	5	✓
26	bugzilla.redhat.com	2 482	1	✓
27	rhn.redhat.com	2 019	4	✓
28	www.mozilla.org	1 785	4	✓
29	crbug.com	1 458	1	✓
30	www.ubuntu.com	1 397	3	✓
		...		
33	bugzilla.mozilla.org	1 075	1	✓
		...		
47	wpscan.com	791	5	✓
		...		
66	medium.com	443	6	✓

We analyze which domains contain suitable descriptions of vulnerabilities and to what extent they are usable. Based on their characteristics, we derive groups of references. In the following, we present and describe the six identified groups in conjunction with some sample domains.

1. **Version control and bug tracker services**
   Examples: GitHub, crbug.com, bugzilla.mozilla.com
   These sites mostly contain program code, output and log files, technical descriptions, and bug discussions. A more abstract description of the vulnerabilities is rarely found. The domains are operated by the producers of the software, but contributions by users are also possible. Hence, there is not always an information verification by experts. On some sites the structure of the references is always identical, on others the structure is inconsistent.
2. **Mailing Lists**
   Examples: lists.fedoraproject.org, lists.apache.org, lists.debian
   Contributions origin from different individual users and are mostly unstructured

and inconsistent texts and code fragments. As a result, some references to a domain may allow a unique mapping from CVE-ID to text, while this is not possible for other references to the same domain. Descriptions of vulnerabilities may be present, however, these are predominantly technical details. On some domains, vulnerabilities fixed with an update are also only mentioned without further text.

3. **Patchnotes**
   Examples: support.apple.com, oracle.com, helpx.adobe.com
   These are often maintained large commercial vendors. A single reference to one of the domains in this group typically contains information about many different vulnerabilities that have been closed with an update. On some domains, descriptions of the vulnerabilities are published, on others, the CVE-ID is only mentioned. References to one and the same domain have mostly identical structures over the whole observed period. The articles are written by employees of the respective companies.
4. **Security Advisories**
   Examples: tools.cisco.com, security.gentoo.org, ibm.com
   Vendors describe vulnerabilities in their own products in more detail on domains in this group. Often, only one vulnerability is covered in a reference. The structures of the references on a domain are the same. The descriptions of vulnerabilities are relatively detailed. The authors are employees of the respective companies.
5. **Third party articles about vulnerabilities**
   Examples: wpscan.com, zerodayinitiative.com, packetstormsecurity.com
   Companies or users publish articles on domains of this group about weak points in the products of other manufacturers. In some cases, this is part of a commercial business model based on services. Unlike the vulnerability-focused mailing lists, the structure of these posts is consistent. Some sites publish contributions written by professional employees, while on other sites host articles by unverified authors.
6. **Blog posts and social media**
   Examples: medium.com, twitter.com, groups.google.com
   References to domains from this group show high diversity. The structure of the contributions is inconsistent. Authors may be professional contributors as well as unverified users. A clear assignment of CVE-ID to text depends on the authors of the specific contributions, not on the website itself.

Our criteria for the usability of texts (*cf.* Sect. 12.3) cannot be met by general purpose crawling approaches, like trafilatura [369], which ignore the characteristics of the target-domain. Instead, solutions must be tailored to the target domain. This is the only way to extract texts from the references that meet our requirements. Since numerous domains are referenced, a pre-selection must be made.

The presented groups differ in terms of the usability of the references. Within the groups the domains are differently suitable. Ideally, text references' enable a unique mapping from a CVE-ID to text. The text must be an abstract description, since technical details such as code descriptions out-of-scope in the present work. Since web scraper use the Hypertext Markup Language (HTML) source code's structure of the domain to extract the correct text, individual references to a domain should therefore always have the same structure.

A unified structure is used on domains where contributions are published or at least reviewed by a single entity. For the first and second group, there is only a higher-level structure, but not a uniform structure of the actual contribution. For example, the basic structure of a reference to an issue in GitHub is always the same, however, the structure of the actual issue description might differ in each case. Table 12.2 shows a simplified overview of the different groups, whether they meet the uniqueness-, uniformity-, and abstract-text-requirements based on a 5-point scale.

Table 12.2 Overview of the 5-point scale evaluation for the usability of the different groups in combination with the usual origin of the content

Group	Origin	Unique	Uniform	Abs. text
VCS/Bug Tracker	*User*	✓✓✓	✓✓✓	✓✓
Mailing Lists	*User*	✓✓	✓	✓✓
Patchnotes	*Vendor*	✓✓✓	✓✓✓✓✓	✓✓
Advisories	*Vendor*	✓✓✓✓✓	✓✓✓✓✓	✓✓✓✓
Third Party	*3rd-P.*	✓✓✓✓	✓✓✓✓✓	✓✓✓✓
Blogs/Social Media	*User*	✓✓	✓	✓✓✓

Domain Selection For the domain selection, it must be considered whether it is worth the effort to adapt a web scraper for a domain. Pages with the same structure and content require less effort and promise a better yield, as the texts will be more likely to meet the established criteria.

Starting from the frequency ranking of domains (*cf.* Table 12.1), a domain selection is made based on the domains group and the group ranking of Table 12.2.

- ibm.com
  Group 4–3 447 References
  IBM publishes collected information about vulnerabilities in its own products. The individual texts are rather short. The assignment of text to CVE ID is easy thanks to the uniform structure of the articles.
- tools.cisco.com
  Group 4–3 019 references.
  Cisco publishes detailed descriptions for vulnerabilities in its own products or in third-party products that Cisco uses or integrates into its own products, such as frameworks. In addition, technical details and code are sometimes included. The structure of the articles is very similar.
- zerodayinitiative.com
  Group 5–2 899 references.
  Trend Micro[9] acts as a middleman between the discoverers of zero-day vulnerabilities and the manufacturers of the affected products. The advisories are then published. The structure and type of description are always the same.
- talosintelligence.com
  Group 5–1 335 References
  Talos is a commercial company belonging to Cisco offering services and products related to IT security. The website publishes articles about vulnerabilities discovered by Talos. The articles are very detailed. The text on the website includes code, version numbers, CVSS vector and other information in addition to the description. However, the text itself is structured by headings that are consistent for all posts.
- qualcomm.com
  Group 3/4–1 048 References
  Contains information collected monthly on vulnerabilities in Qualcomm products. Descriptions are brief. The structure is consistent, and the articles are sorted into tables. Partially the URLs deposited in the NVD are incorrect, because Qualcomm has changed the Uniform Resource Locator (URL) scheme over time. However, the monthly posts are still accessible under a modified URL.
- support.f5.com
  Group 5–932 references
  F5 provides commercial IT security services and products. The referenced papers

[9] website: trendmicro.com/de_en/business.html

describe individual vulnerabilities in products developed by F5. The structure is consistent.

- wpscan.com
  Group 5–803 references.
  A provider that rehashes vulnerabilities from the WordPress ecosystem and offers services related to the security of WordPress installations. For each CVE Identification number exists a short description, the structure of the page is the same throughout.
- intel.com
  Group 4–771 references
  Intel publishes here lists of vulnerabilities that have been fixed with an update. The structure of the pages is always identical, and an assignment is possible without any problems.
- snyk.io
  Group 5–671 references
  Snyk offers several commercial vulnerability management products. The company maintains a public database of vulnerabilities in Open-Source Software (OSS), respectively in open source ecosystems like Node Package Manager (NPM) or Maven. The descriptions are sometimes very detailed and the structure of the contributions is always identical.

The selected web pages are referenced a total of 14 925 times. However, it is to be expected that not all references are available anymore.

Since we use novel information sources for our proposal, we also want to verify, whether information sources, which are currently regarded as high-quality information sources, *e.g.*, exploit-DB and Common Weakness Enumeration (CWE), are able to improve current models. Hence, we select their texts as well to train a *ground-truth* model, to identify, whether additional data have an actual influence in the training process.

Special Features of Twitter Twitter is an important medium in IT security and has been the subject of several works [125, 175]. Twitter is also frequently referenced in NVD and is found among the 100 most referenced websites. However, a preliminary analysis shows that the references are unusable. In some cases, only user profiles are referenced, such as for CVE-2021-25179[10]. The reference twitter.com/gm4tr1x is

[10] https://nvd.nist.gov/vuln/detail/CVE-2021-25179

the profile of the vulnerability's discoverer[11]. User profiles provide no meaningful information for the present work. Generally, such references are not in line with the CVE's reference requirements[12]. In our dataset 17% of references on Twitter are links to profiles. Other references are retweets, such as seen in CVE-2021-27549[13], yielding the same problem. The original tweet is also referenced in the NVD.

Some Twitter references actually contain a description of the vulnerability. Twitter is thus very important as a medium to exchange information between experts in a short amount of time, but cannot serve well as a source for texts in this work.

12.4 Implementation

While the previous section (*cf.* Sect. 12.3) examined the space of available references and accompanying requirements, this section explains the process of web scraping and model training.

12.4.1 Web Scraping

The selected domains (*cf.* Sect. 8) are publicly available, but no API exists to retrieve their content. So the texts have to be extracted from the pages via web scraping.

Through the robots.txt[14], the operator of a website can select which bots should access which URLs. However, this employs only a soft restriction, since it cannot be technically enforced. With the Python library `urllib`, the robots.txt of the selected domain is checked whether access to the NVD referenced in the URLs is allowed. In some cases, a delay between requests is desired due to the non-standard directive *crawl-delay*. The developed web scrapers respect this accordingly.

While trafilatura [369] seem promising, our insights from Sect. 12.3 show, that it should be avoided in the present work. Fig. 12.3 shows an example of the relevant part of Cisco's website. It contains the requested description as well as other texts that is present on this page. The static texts, such as headings and various legal information, are the same for each reference and represent noise. While trafilatura

[11] The reference to the SolarWinds vendor page lists the name *Gabriele Gristina* as the discoverer. His LinkedIn and GitHub account are also referenced, in addition to the Twitter profile.

[12] https://www.cve.org/ResourcesSupport/AllResources/CNARules#section_8-3_cve_record_reference_requirements

[13] Referencing https://twitter.com/0xabc0/status/1363855602477387783

[14] https://www.robotstxt.org/

removes parts such as the page header, bigger chucks like the legal information are still present during text extraction.

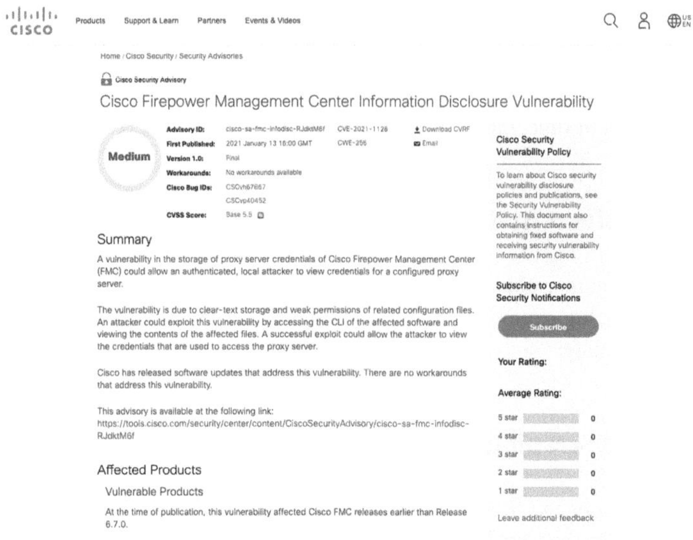

Fig. 12.3 Example of Cisco website, referenced in https://nvd.nist.gov/vuln/detail/CVE-2021-1148

The results are similar for *ibm.com*, *zerodayinitiative.com*, *wpscan.com*, *talos-intelligence.com*, and *snyk.io*. Some unwanted content could still be removed by filtering the output by trafilatura, but this would require post-processing, which negates the idea of trafilatura. On some pages of *qualcomm.com* and *intel.com* multiple vulnerabilities are treated together, which introduces noise in the training process. During implementation we identified, that, *e.g.*, *qualcomm.com* changed its URL structure, so that some of the referenced URLs are unavailable. However, a manual search shows that the pages themselves are still present under other URLs.

Since Trafilatura cannot execute JavaScript, the pages of *support.f5.com* cannot be retrieved at all. This is because the server responds to initial Hyper Text Transfer Protocol (HTTP) GET requests for the referenced URLs with a JavaScript file embedded in HTML. In a browser, the script is then executed and thus the actual page content is loaded. While Trafilatura might work in other contexts, it is, in many ways, not suitable for the present work, partly due to the special requirements (*cf.* Sect. 12.3).

Several other technologies offer better controllability and in-depth filtering capabilities. Selenium[15] is a framework for automated testing of web applications and enables automatic control of full-featured web browsers in the background, *e.g.*, Google Chrome and Mozilla Firefox. Through application programming interfaces (APIs) for various programming languages, including Python, the web browser can be controlled. The APIs allow access to the Document Object Model (DOM) representation of the HTML content of the accessed web page. For testing, user interaction can be simulated, such as clicks or input. Selenium thus provides everything necessary to JavaScript enriched web pages. However, it is not a lightweight and particularly fast solution.

Beautiful Soup[16] is an OSS web scraping library for Python. It allows parsing of HTML files. The user can navigate through the API structure to get selected parts of the web page. Beautiful Soup is lightweight and faster than Selenium, but is limited to HTML content. If parts of the page are reloaded using JavaScript, Beautiful Soup cannot access them accordingly.

Since the amount of references to be retrieved with the Web Scraper is limited to 14 925 and the retrieval is done only once, time plays only a minor role. Rendering the web pages with Selenium takes most of the time. The speed can be increased linearly by parallelization.

The program is structured according to the producer-consumer design pattern. First, all URLs are collected, then multiple threads are started to process the URLs in parallel. The correct web scraper is selected based on the URL.

The web scrapers for *talosintelligence.com* and *intel.com* are implemented using Beautiful Soup, and Selenium is used for the rest of the pages. The Beautiful Soup based web scrapers take about a second to retrieve and parse a web page, while Selenium based web scraper usually takes about five seconds. The web scraper first waits until the requested page is fully loaded and no more JavaScript is executed. Sometimes this leads to a blockade, because JavaScript is executed permanently. Therefore, the execution times out after 20 seconds. The page with the actual text is usually fully loaded by that time and can be parsed. Since such timeouts seldom occur, resulting idle times are negligible. In total, a complete run over all references in the selection took about 12 hours at a measured Internet speed of about 50 MBit/s and five parallel web scrapers.

---

[15] https://www.selenium.dev/

[16] https://www.crummy.com/software/BeautifulSoup/

As mentioned before, some URLs for *qualcomm.com* are unavailable. Hence, the web scraper is set up to first check the NVD's reference and if it fails, start another attempt corrected URL, corresponding to the current URL scheme. This reliably fixes the URL problems for *qualcomm.com*.

Further, we use known high quality information sources, which relate their texts to the NVD's vulnerability identifier to evaluate, whether the quality of the information sources might be problematic. Here we use the CWE[17] and exploit-DB[18] as information sources. As for the CWE we add the texts of the referenced CVE id to the training set for a *quality assurance model* (*cf.* Table 12.6), while for exploit-DB we collected the texts for 2 402 vulnerabilities, based on a listing by Mitre[19] (Table 12.3).

Table 12.3 Number and proportion of each reference successfully retrieved by the web scraper from the preselection

	References	
Webpage	Crawled	Ratio
ibm.com	2 868	0.83
tools.cisco.com	3 004	0.99
zerodayinitiative.com	2 899	1.0
talosintelligence.com	1 201	0.89
qualcomm.com	697	0.66
support.f5.com	740	0.79
wpscan.com	35	0.04
intel.com	731	0.94
snyk.io	627	0.93
Total	12 802	0.85

In total, 12 802 references (85%) of the 14 925 original ones are successfully retrieved. During the crawling process, we identified problems with *wpscan.com*. The domain permits all bot access in its robots.txt, but blocks all requests after five initial ones in quick succession. This means that it is not possible to retrieve many references in a meaningful way. Of the 803 references originally available, only 35 were retrieved.

17 https://cwe.mitre.org/

18 https://www.exploit-db.com/

19 https://cve.mitre.org/data/refs/refmap/source-EXPLOIT-DB.html

Fig. 12.4 shows the distribution of the lengths of the successfully retrieved texts. The average is 817.11, and the median is 532 characters. The texts are between 32 and 32 206 characters long. Thus, the obtained texts are significantly longer than the descriptions from NVD (*cf.* Fig. 12.1).

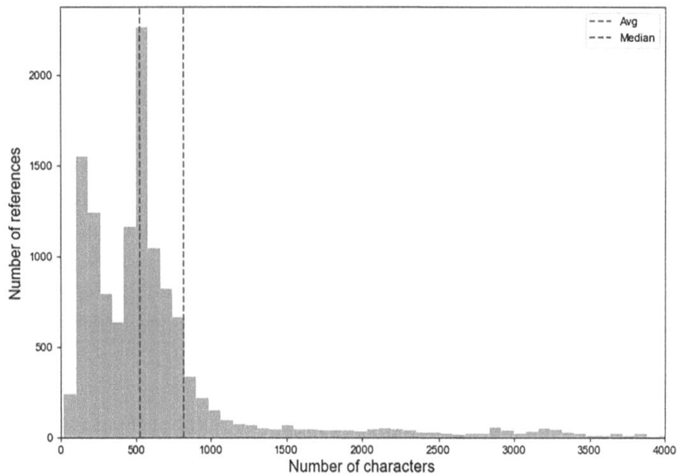

Fig. 12.4 Distribution of text lengths retrieved with the web scraper

Fig. 12.4 shows the distribution of lengths of successfully retrieved texts. The average is 817.11, the median is 532 characters. The texts are between 32 and 32 206 characters long. Thus, the obtained texts are significantly longer than the descriptions from NVD (*cf.* Fig. 12.1).

12.4.2 Deep Learning Classifier

The goal of the work is to predict the entire CVSS basis vector, the problem is split into several sub-problems in the form of classifying the individual components of the vector. The components of the CVSS vector are Attack Vector (AV), attack complexity (AC), privileges required (PR), user interaction (UI), Scope (S), Confidentiality Impact (C), Integrity Impact (I), and Availability Impact (A)[20]. For

[20] https://www.first.org/cvss/specification-document

each component there is an independent classifier. As a result, eight models must be trained separately.

12.4.2.1 Model Selection

Shahid and Debar [354] use a model based on BERT [47] for their work. A classifier in the form of a fully-connected feed-forward NN is placed on top of the BERT base model in each case. Shahid and Debar [354] use BERT-small [370], rather than the original version of BERT [47]. This model achieves a similar result in various benchmarks with significantly fewer parameters than BERT, but is faster to train. Since eight models must be trained, we adapt this idea to use one of the smaller BERT models. DistilBERT [215] gives an even slightly better results than BERT-small [370] while also having fewer parameters than the original BERT.

For our implementation, the OSS library transformers[21] from Huggingface [371] is used. This provides an abstraction of the actual PyTorch[22] models and provides easy access to many different pre-trained models. DistilBERT [215], BERT-small and BERT-medium[23] [370], among others, are available via the transfomers API.

12.4.2.2 Training

The entire dataset is composed of descriptions from the NVD and texts retrieved from the selected domains (*cf.* Sect. 12.4.1). We crawled 88 979 NVD descriptions and 12 755 texts, for a total of 101 734 data points. In the following, NVD descriptions and retrieved texts are treated identically, *i.e.*, the origin of texts is ignored. The dataset is split into a training set with 75% and a test set with 25% of the texts. This ensures that texts referring to the CVE ID are always also in the same set.

For our *quality model* (called $\text{DistilBERT}_{\text{quality}}$) which uses exploit-DB and CWE as additional data sources, we aim to train a model on a dataset similar to the one described, the number of data points stays the same as before, but the text from exploit-DB and CWE are appended to the training set texts to enhance their quality. During manual analysis, we saw, that this might push the boundary of the 512 tokens, which can be used as input for our language models, but the latter parts of the exploid-DB texts usually contain source code, which should be ignored either way. This differs from the mentioned method of creating a data point for each single text, ignoring its origin. Our motivation behind doing this is (i) when creating a single data point for each single text, including the CWE, we might get different texts all mapping to the same CVSS value, which should worsen the model quality

[21] https://huggingface.co/docs/transformers/index

[22] https://pytorch.org/

[23] https://huggingface.co/prajjwal1/bert-medium

and (ii) this would result in different test sets, making a comparison of model quality difficult.

The training of the individual DistilBERT, BERT-small, and BERT-medium models is performed independently on the Lichtenberg high-performance computer. It provides Graphics Processing Units (GPUs) of type Nvidia Ampere 100 and Volta 100. The batch size is set based on the available GPU. Table 12.4 shows the possible batch size and time needed for six epochs of training including evaluation after each epoch. As one be seen, the training time does not decrease quite linearly with batch size. The speed of GPUs also plays an important role. In experiments, the training could also be performed on Nvidia T40 and K80 with 16Gb memory. Shahid and Debar [354] freeze the layers of the BERT model for the first three epochs of training and only let the classifier adapt.

Table 12.4 Batch size and training duration of the different models (Dist. = DistilBERT, Sm. = BERT-small, Med. = BERT-medium) depending on the used GPUs.

GPU info		Batch size			Time [min]		
Model	Mem.	Dist.	Sm.	Med.	Dist.	Sm.	Med.
A100	40Gb	48	128	56	60	25	35
V100	32Gb	40	96	48	132	50	95
T40/K80	16Gb	24	64	28	–	–	–

12.5 Evaluation

The previously trained models are evaluated in this section. For this purpose, different metrics for the individual classifiers are considered and compared, including white-box indicators to reconstruct the decision process of our models. Finally, we determine whether the additional texts have an impact on the overall score.

12.5.1 Classifier

Table 12.5 shows various metrics (Accuracy, Recall, Precision, F1, Cohen κ) of our classifiers. The F1 scores are arithmetic means (macro weighted), so the different distribution of target variables is not taken into account.

Table 12.5 Accuracy (Acc), Recall (Rec), Precision (Prec), F1 Score, and Cohen's κ (Cohen) for the eight classifiers (Attack Vector (AV), attack complexity (AC), privileges required (PR), user interaction (UI), Scope (S), Confidentiality Impact (C), Integrity Impact (I)) of each model. Calculated on the test set

	DistilBERT					BERT-small					BERT-medium				
	Acc	Rec	Prec	F1	Cohen	Acc	Rec	Prec	F1	Cohen	Acc	Rec	Prec	F1	Cohen
AV	0.93	0.82	0.85	0.84	0.84	0.91	0.82	0.82	0.82	0.81	0.93	0.83	0.85	0.84	0.82
AC	0.96	0.78	0.85	0.82	0.64	0.96	0.8	0.84	0.83	0.61	0.94	0.79	0.84	0.8	0.61
PR	0.87	0.8	0.81	0.8	0.74	0.86	0.8	0.81	0.8	0.73	0.87	0.8	0.82	0.8	0.74
UI	0.95	0.94	0.93	0.93	0.87	0.93	0.94	0.94	0.94	0.88	0.94	0.94	0.94	0.94	0.88
S	0.95	0.92	0.93	0.92	0.85	0.95	0.91	0.93	0.92	0.85	0.95	0.92	0.93	0.93	0.86
C	0.89	0.85	0.86	0.87	0.8	0.89	0.85	0.86	0.87	0.8	0.88	0.85	0.87	0.86	0.8
I	0.9	0.87	0.9	0.89	0.83	0.88	0.87	0.89	0.87	0.81	0.88	0.87	0.89	0.88	0.83
A	0.9	0.75	0.8	0.77	0.81	0.9	0.72	0.8	0.75	0.8	0.9	0.76	0.77	0.76	0.81

All models, except the Availability Impact (A) model, achieve F1-scores above 0.8 for all components. The quality of our classifiers is thus comparable to the classifiers of Shahid and Debar [354]. However, a clear improvement cannot be seen from the metrics.

For AV, all models achieve very good predictions for the overrepresented values *N* and *L*. Although the values *P* and *A* occur very rarely, the classifiers still manage to correctly detect over 70%.

The classifiers for AC, PR, and A work for frequent values, but are much worse for less frequent ones. While the F1 score for AC and PR is unremarkable in each case, this imbalance is evident in the lower Cohen's κ. In particular, for AC *H*, the classifiers are not reliable in this way. For A, only around 40% are correctly detected for *L*, which is the lowest rate of all classifiers.

For UI, S, C, and I, the classifiers are good to very good, with the best results for UI.

Overall, a highly uneven distribution of values in the dataset tends to lead to worse results in predicting the underrepresented values, which is a common problem with DL.

12.5.2 CVSS Score

To obtain the total CVSS score, the results of the classifiers of a model are combined. From the individual components, the score is calculated according to the CVSS standard[24]. The obtained scores are compared with the expert generated scores in the NVD.

Table 12.6 shows the Mean Absolute Error (MAE) and MSE and the fraction of vulnerabilities where the predicted score is higher, lower, or equal to the NVD's Ground Truth (GT). We also included the results of our *high-quality references* model, which was trained on NVD, CWE, and exploit-DB texts (*cf.* Sect. 12.4.1 and 12.4.2.2), as well as a comparison of our approach against the proposals by Shahid and Debar [354] and Spanos et al. [360]. In Fig. 12.5 the distribution of differences from true to predicted score of DistilBERT classifiers is shown. The average difference is 0.6, while 75% of all predictions are within the range of 1 around the actual score.

Table 12.6 Mean Squared Error (MSE), Mean Absolute Error (MAE), and proportion of correct ($Pred_c$), too high ($Pred_h$), and too low ($Pred_l$) predictions

	MSE	MAE	$Pred_c$ [%]	$Pred_h$ [%]	$Pred_l$ [%]
DistilBERT	1.44	0.61	62.1	20.5	17.3
$\text{DistilBERT}_{\text{quality}}$	1.44	0.61	62.2	17.5	20.2
BERT-small	1.52	0.624	61.6	20	18.2
BERT-medium	1.47	0.617	61.6	20	18.1
[354]	1.79	0.73	55.3	-	-
[360]	-	1.74	-	-	-

Fig. 12.6 shows the ratio of true to predicted scores for the DistilBERT classifiers. All models predict scores rather higher than lower compared to the original score, but the difference is very small. In general, it is better to score a vulnerability too high than too low, but depending on the use case, this can be a problem (*e.g.*, if an information overload is already present). Table 12.6 shows, that there is no difference, in terms of MSE and MAE, compared to the reference OSINT texts. This is either the result of a negligible influence of the concatinated text parts during the training and classification process (*cf.* Sect. 12.4.2.2), or the already high quality of other OSINT references. Either case should be inspected in future work. Another striking phenomenon is the series of predictions with a score of 0.0. These are 111 (DistilBERT), 151 (Bert-small), and 121 (Bert-medium) predictions. If N is predicted for each of the C, I, and A components, this makes the Impact Sub Score (ISS)[25] equal to 0. The ISS is multiplied by the other CVSS components, resulting in 0.0 for the total score. There are no vulnerabilities that actually have

[24] https://www.first.org/cvss/specification-document

[25] https://www.first.org/cvss/specification-document

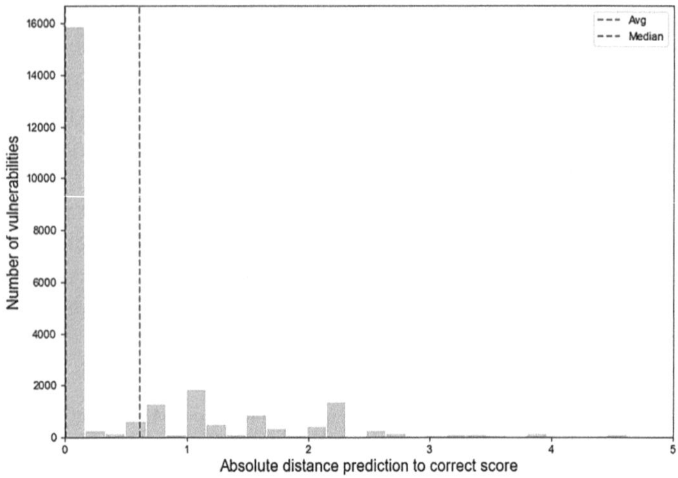

Fig. 12.5 Distribution of difference between the predicted score with the DistilBERT classifiers and Ground Truth (GT) Common Vulnerability Scoring System (CVSS) score

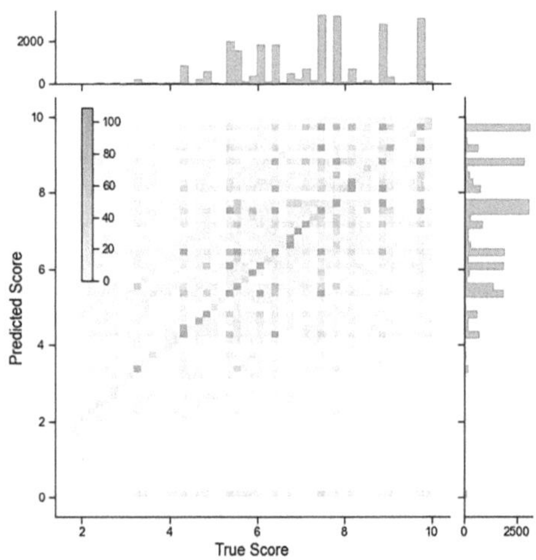

Fig. 12.6 Distribution of DistilBERT classifier predicted scores and the correct Common Vulnerability Scoring System (CVSS) scores

this combination. Since there are few such predictions, the problem goes unnoticed in the metrics. However, critical vulnerabilities (with scores above 9.0) would be directly discarded because of this problem.

12.5.3 Explainability and Interpretability

For IT security applications, it is important that ML procedures are explainable. DL models usually lack this property. It is difficult to understand what the model has learned in its entirety. However, individual examples provide some insight into the model.

The PyTorch library *Captum*[26] implements various algorithms that help explain DL models. For the following examples, we use Layer Integrated Gradients from Captum on the trained DistilBERT classifiers [372]. Table 12.7 shows example description results for the CVSS prediction of CVE-2016-0775[27]. Words that argue

Table 12.7 Influence of words on prediction of Attack Vector (AV) (with the parameters N – Network, L – Local, A – Adjacent, P – Physical) based on description of CVE-2016-0775. Green corresponds to a positive influence, red to a negative influence. Predicted AV: *L*. Correct AV: *N*

AV	Text
N	Buffer over flow in the ImagingFliDecode function in libImaging/Fl iDe code.c in Pillow before 3.1.1 allows remote attackers to cause a denial of service (crash) via a crafted FLI file .
L	Buffer overflow in the Imaging FliDecode function in libImaging/ Fl iDe code .c in Pillow before 3.1.1 allows remote attackers to cause a denial of service (crash) via a crafted FL I file .
A	Buffer overflow in the ImagingFliDecode function in libImaging/Fl iDe code.c in Pillow before 3.1.1 allows remote attackers to cause a denial of service (crash) via a crafted FLI file .
P	Buffer overflow in the ImagingFliDecode function in libImaging/Fl iDe code.c in Pillow before 3.1.1 allows remote attackers to cause a denial of service (crash) via a crafted FLI file .

[26] https://captum.ai/

[27] https://nvd.nist.gov/vuln/detail/CVE-2016-0775

for the classification in each class are marked in green, words that argue against are marked in red. For clarity, only the words with the greatest influence are marked. The classifier's prediction is *L*, but *N* would have been correct.

For human experts, the phrase "[..] allows remote attackers to [..]" is a clear indication that the Attack Vector (AV) is *Network*. As Table 12.7 shows, "remote attackers" also argues for *N*. However, the word "file" at the end of the description is strongly scored against *N* and for *L*.

This way, a rough understanding of the representation learned by the models is gained. Many assessments are reasonable, but the DL models still remain a black box for users.

12.5.4 Influence of Additional Texts

We evaluate whether the texts retrieved via web scrapers have a positive effect compared to using the NVD's descriptions only. For this purpose, new DistilBERT models are trained on the descriptive texts from NVD only. DistilBERT was chosen because the models performed best overall in our previous evaluation (*cf.* Table 12.5). The models trained exclusively on NVD descriptions are called DistilBERT$_{desc}$ in the following.

Table 12.8 shows the results of the evaluation on different datasets. The *Combined* dataset is the previously used dataset of descriptions (2016–2021) and retrieved texts. *Desc* contains only the descriptions, but no additional reference texts. A new test dataset *Desc*$_{2022}$ is used, consisting of 5 641 descriptions published between January and May 2022. These descriptions were not previously used for training and evaluation.

Table 12.8 Mean Squared Error (MSE) and Mean Absolute Error (MAE) on different datasets.

	Combined		Desc		Desc$_{2022}$	
	MSE	MAE	MSE	MAE	MSE	MAE
istilBERT	1.44	0.61	0.455	0.203	1.941	0.811
DistilBERT$_{desc}$	0.544	0.248	1.393	0.604	1.928	0.788

On the combined dataset, the DistilBERT$_{desc}$ classifiers achieve significantly better scores. Over 80% of the predictions were correct. We saw significant improvements for AV, AC, and A over the combined trained DistilBERT. On the NVD descriptions, the combined trained DistilBERT is significantly better than DistilBERT$_{desc}$. The additional reference texts do have a positive effect. For $Desc_{2022}$ the models are on par. DistilBERT$_{desc}$ is only slightly better here.

These results are rather unexpected, *i.e.*, the model, which is trained purely on NVD descriptions performs significantly better on the prediction of all texts (including references) and the other way around, and both models perform similar on a new dataset. The performance of the DistilBERT model trained on the combined dataset can be explained with the higher robustness, but the former cannot. The only clue might be the high quality of the NVD descriptions, but this is counterintuitive to the results of Kuehn et al. [5].

However, since all models classify texts, which is comparable to predicting a discrete value, small differences might lead to a large difference in the score. If, for example, the models are tasked to predict the CVSS score for CVE-2022-23442[28] and it would falsely predict the C impact as *None*, the impact score would be 0 and with it the whole CVSS score would result in 0. There may be a small but crucial difference between the DistilBERT-combined and DistilBERT-descriptions classifiers for a single CVSS component, but the cause of the surprising results from Table 12.8 could not be determined.

12.6 Discussion

This section discusses the results (*cf.* Sect. 12.5) and points out future work.

Analysis of References and Web Scraping The preliminary analysis identified sources of textual vulnerability information besides the ones in the NVD (**RQ1**). Hereby, we grouped sources and rated their vulnerability uniqueness, uniformity of texts, and the presence of an abstract vulnerability description (*cf.* Table 12.2). Due to the strict selection, only references from groups 3, 4 and 5 (*cf.* Sect. 12.3.2) are eligible. However, the retrieved texts for this purpose contain almost exclusively the abstract description. Whether noise would play a major role in the texts is unclear, which could be explored for further work. While relaxing our criteria would make significantly more web pages usable, the current used language model might not be suited for such task.

Since the used DL models are optimized for natural language, log files and source code could not be used. Some of the references mix code and natural language. The

[28] https://nvd.nist.gov/vuln/detail/CVE-2022-23442, with an expert rated CVSS vector of *AV:N/AC:L/PR:L/UI:N/S:U/C:L/I:N/A:N*

currently available Natural Language Processing (NLP) models are not able to use source code in addition to natural language. Separate models for source code could be used for this in the future. With such an improvement, future work can build on our reference analysis (*cf.* Sect. 12.3.2) and try to gather groups with mixed information types.

But, with the current state of research, adaptation to each website is necessary, which increases the effort linearly with the number of websites to create large datasets. Technically, there is otherwise little potential for optimizations to the implementation of web scraping. Web scrapers can be parallelized as is and used productively. Other solutions not based on manually customized web scrapers are not currently available. The need for manually adapted web scraping would also be eliminated by a uniform standard, *e.g.*, Common Security Advisory Framework (CSAF)[29].

Deep Learning Classifier Several different current DL models were successfully trained and evaluated as classifiers for the components of CVSS vectors.

The retrieved reference texts could be used as a dataset together with the descriptions. The obtained classifiers achieve good scores in several metrics [224, 354]. In particular, the DistilBERT model provides good results. Therefore, texts from OSINT sources are usable for CVSS prediction, but do not have a clear positive impact on the result in this form **RQ2**. It is possible to use OSINT as a textual source as a basis for CVSS prediction. Since the models require little time to train, it would also be possible to train regularly to incorporate new information into the classifier's decision. However, the expected positive effect on the quality of the models did not occur due to the additional texts (*cf.* Sect. 12.5). Even just including known high quality OSINT sources besides the NVD did not positively influence the trained models. Either, the models are not influenced by these texts anymore due to the high bias of the previous text passages, *i.e.*, the models are already saturated by the texts, or the texts used in the other training processes are of very high quality already. We omitted using additional high quality sources like ATT&CK, due to the given results (not any movement of results despite the additional texts).

Usability in Real-World Applications The results in Table 12.8 show, that our DistilBERT model, trained on the combined dataset, is able to predict the correct vulnerability score based on the description with a MSE of 0.46 and MAE of 0.2. This very accurate compared to other currently available methods Table 12.6. Whether our results are suitable to support or even replace human practitioners in

[29] https://oasis-open.github.io/csaf-documentation/

cybersecurity depends on the use case. If the goal is to overcome current information overflow and get a first hint of the vulnerability landscape, it should be considered using a tool guided by the proposed CVSS models. The accuracy of predictions is very good and enables to get a hint of the final CVSS score. If the use case should be completely outsourcing the NVD expert knowledge to such models, we would not recommend this step. ML models can only be as good as their training data and might be biased or be prone to domain shift, *i.e.*, perform worse on new data due to drifting terminology. While the latter problem might be solved by regular retraining steps of the models, the former cannot simply be solved. One might, however, argue, that the bias can also be given by human personnel, but quality assurance steps in the pipeline of vulnerability assessment should always consider such problems, *e.g.*, by verifying scores by different personnel. The models could, however, support NVD personnel in the assessment process of new vulnerabilities and the training of new personnel. Further, should the insights of human personnel in the IT infrastructure always outweighs the prediction of singular models.

Limitations & Future Work In the area of web scraping, the paper is limited by the structure of the referenced web pages (*cf.* Sect. 12.3). Future work may simplify web text collection. This could lower the effort required to adapt web scrapers to different web pages. Optimally, a solution would be as easy to use as Trafilatura, while still being able to find only the text related to a specific CVE ID. Also, a Graphical User Interface (GUI) based program could be developed that allows the selection of elements on a web page. Based on this selection, the program could then generate the necessary code for the web scraper in a selected web scraping framework.

For the CVSS classifier it needs to be investigated whether more texts lead to better results. The influence of noise should be clarified as well and, based on this, the criteria for usable text established in this work should be evaluated again.

The existing method may have potential for optimization at various points. Depending on the specific use case, all texts from references could be used for training. The results in Table 12.6 and Table 12.8 suggest that there might be potential for a currently unknown combination of high quality texts and different volumes of texts, that could improve the quality of CVSS prediction models. Unrealistic scores with a score of 0 can be prevented by minor additions to the logic of the classifiers. Instead of BERT models trained on general language, models trained specifically on texts from IT security could also serve as a basis.

Data augmentation can be used to improve or compensate for the uneven distribution of different variables in the dataset [203]. Sect. 12.5.3 has shown that the decisions of the classifiers are only partially understandable. Further work can improve the explainability and interpretability of the models.

Additionally, the present work lacks a comprehensive comparison against previous work. This results either from missing published models of previous work to reconstruct the results for a fixed test-set or from the disjoint metrics, that are published. Hence, a comparison was not possible.

12.7 Conclusion

Vulnerabilities in IT systems pose a major threat to society, businesses, and individuals. A fast and reliable assessment of newly published vulnerabilities is therefore necessary. The increasing amount of new vulnerabilities makes timely assessment by human experts difficult. Therefore, various works [5] [224, 354] deal with automated prediction of CVSS vector, score or level by ML. These works, however, focused on the NVD data alone, rather than using additional OSINT texts for vulnerabilities. In this work, the possibility of using OSINT as a vulnerability texts source was investigated. First, a preliminary analysis of the referenced domain in NVD vulnerability entries was performed. In this, the domains are classified into groups based on criteria according to their usability. This resulted in a pre-selection of domains which later are scraped. The reference texts and NVD descriptions, as well as selected high quality sources, were used as training set for different DL-based classifiers. Finally, the classifiers were evaluated and their quality was assessed. The classifiers achieve good results in predicting the individual components of the CVSS vectors. The CVSS scores computed from them have low error rates and real world usage of the proposed models should be considered for certain use cases.

Reducing Information Overload: Because Even Security Experts Need to Blink 13

Abstract

Computer Emergency Response Teams (CERTs) face increasing challenges processing the growing volume of security-related information. Daily manual analysis of threat reports, security advisories, and vulnerability announcements leads to information overload, contributing to burnout and attrition among security professionals. This work evaluates 196 combinations of clustering algorithms and embedding models across five security-related datasets to identify optimal approaches for automated information consolidation. We demonstrate that clustering can reduce information processing requirements by over 90% while maintaining semantic coherence, with deep clustering achieving homogeneity of 0.88 for security bug report (SBR) and partition-based clustering reaching 0.51 for advisory data. Our solution requires minimal configuration, preserves all data points, and processes new information within five minutes on consumer hardware. The findings suggest that clustering approaches can significantly enhance Computer Emergency Response Team (CERT) operational efficiency, potentially saving over 3750 work hours annually per analyst while maintaining analytical integrity. However, complex threat reports require careful parameter tuning to achieve acceptable performance, indicating areas for future optimization. The

Minor changes regarding language, style, and formatting have been made in this chapter to provide a consistent experience.

Supplementary Information The online version contains supplementary material available at https://doi.org/10.1007/978-3-658-49241-0_13.

P. Kühn, *Proactive Cyber Threat Intelligence*, Technology, Peace and Security | Technologie, Frieden und Sicherheit,
https://doi.org/10.1007/978-3-658-49241-0_13

code is made available at https://github.com/PEASEC/reducing-information-overload.

Original Publication ***Kuehn, P.**, Bayer, M., Frey, T., Kerk, M., Reuter, C., "Reducing Information Overload: Because Even Security Experts Need to Blink," in* Availability, Reliability and Security—ARES 2025 Workshops, *Ghent, Belgium: Springer Nature, 2025.*

| ICORE B[1]

13.1 Introduction

The cybersecurity threat landscape continuously evolves, with attackers deploying increasingly sophisticated tactics while security findings proliferate across multiple channels. Security personnel struggle to process high volumes of textual reports [27], impeding their primary mission of threat identification and infrastructure protection. Despite existing frameworks like the Cyber Threat Intelligence (CTI) cycle [30, 189] and automation methods [5, 224], information processing challenges persist. While CTI—the process of collecting and analyzing security data to derive actionable recommendations—can be aggregated in threat intelligence platforms (TIPs) [373], the diversity of sources and evolving threats creates significant information overload [111].

CERTs, as organizational security incident coordinators [374], require current threat intelligence for effective response. Studies reveal that 45% of CERT teams process only critical reports due to understaffing [375], while 13% lack capacity for new information and 11% cannot manage existing volumes. Recent research [27, 28] reinforces these challenges, with 47.6% of analysts reporting burnout and 46.6% identifying threat monitoring as their most time-consuming task. For 19.2% of analysts, automating threat alert enrichment through incident correlation represents a critical priority [27]. Kaufhold et al. [28] highlight persistent manual processes in technical information exchange, redundancy checks, and general automation needs, underscoring the urgency for enhanced information processing solutions.

Goal This research evaluates clustering algorithms' efficacy in supporting CERT threat information processing. Clustering enables efficient threat analysis by allowing rapid overview of related data points before detailed investigation. We assess various embedding-clustering algorithm combinations against derived requirements,

[1] Updated 5. Februar 2025.

with particular emphasis on threat messages and security advisories from both commercial vendors and security researchers. This investigation addresses our primary research question: *Which cluster algorithm and embedding combination is suitable to reduce CERT personnel's information overload* (**RQ**)?

Contributions This work advances current research through two primary contributions: (i) introduction of THREATREPORT, a novel labeled threat report corpus (**C1**) and (ii) a comprehensive performance comparisons of 14 clustering algorithms on the created embeddings across the five diverse datasets (**C2**).

Outline The remainder of this paper is structured as follows: Sect. 13.2 examines related work and identifies research gaps. Section 13.3 details our methodology, followed by our comprehensive evaluation results in Sect. 13.4. Section 13.5 discusses findings and limitations, while Sect. 13.6 summarizes our contributions.

13.2 Related Work

We present related work in embeddings, clustering algorithms, and evaluation metrics, culminating in the identification of our research gap.

Embeddings Embedding methods transform data points into vector representations where similarity is preserved through spatial proximity. These range from simple word frequency approaches to sophisticated language models encoding semantic relationships [45, 376]. Document-level encoding presents unique challenges for threat intelligence processing.

Traditional approaches include Bag of Words (BoW), which records absolute term frequencies using a global vocabulary [377], and Term Frequency-Inverse Document Frequency (TF-IDF), which weights terms by their document frequency [378]. Recent approaches use Bidirectional Encoder Representations from Transformers (BERT) [47], with Sentence-BERT (SBERT) specifically optimized for embedding longer text units [45]. The *MTEB* benchmark provides comprehensive performance comparisons of differently-sized large language model (LLM), including clustering efficacy [379].

Clustering Clustering algorithms group data points based on similarity metrics such as cosine distance or silhouette scores [378, 380, 381]. Traditional methods range from centroid-based K-Means [382] requiring predefined cluster counts to density-based Density-Based Spatial Clustering of Applications with Noise

(DBSCAN) [383] supporting arbitrary cluster shapes. Recent research explores deep learning approaches that leverage intermediate representations [384–390].

In security contexts, clustering facilitates log summarization [391], Android permission analysis [392], and cybersecurity event detection in social media [11, 223, 393] using various techniques from local sensitivity hashing to neural networks. Vulnerability management benefits from clustering through alternative vulnerability classification [394].

Evaluation Text clustering evaluation employs both internal metrics (assessing compactness and separability without ground truth) [395] and external metrics (requiring labeled data) [378]. Rosenberg and Hirschberg [396] and Amigó et al. [397] highlight limitations of traditional metrics like purity and entropy, particularly for edge cases. The V-measure framework [396] combines homogeneity (cluster label consistency) and completeness (label distribution) metrics, providing comprehensive clustering quality assessment. Recent frameworks [19] integrate multiple algorithms, datasets, and metrics for systematic evaluation.

Research Gap While existing research addresses clustering of security information, it primarily focuses on short-form content (*e.g.*, social media posts) [11, 393] or traditional embedding methods [11, 223]. No comprehensive evaluation exists, which compares modern embedding-clustering combinations for longer security texts, such as security advisories or threat reports. This gap is particularly significant given the increasing volume and complexity of security documentation requiring efficient processing by CERT personnel.

13.3 Methodology

We present the data used in this work and the requirements for document embeddings, clustering algorithms, and evaluation metrics.

13.3.1 Text Corpora

This study employs multiple datasets to evaluate the selected clustering algorithms across three distinct use cases: (I) effectiveness in processing threat-related short messages and threat reports, (II) performance in handling SBR across diverse products, and (III) comparative analysis on non-security short messages. Exemplar texts from each corpus are presented in Listing 13.1, while Table 13.1 provides a comprehensive overview of the datasets' structural characteristics.

Table 13.1 This table outlines the structural information of the datasets $c \in [CySecAlert, MSE, ThreatReport, SBR, SMS]$. L_c is the sequence $len(dp_i)$ in character for all $dp_i \in c$. It shows the size $|L_c|$, the average length $\overline{L_c}$, the median $\widetilde{L_c}$, the minimum $\lfloor L_c \rfloor$ and maximum $\lceil L_c \rceil$ data point length, and the number of ground truth clusters ($\#L_c$) of c

c	$\|L_c\|$	$\overline{L_c}$	$\widetilde{L_c}$	$\lfloor L_c \rfloor$	$\lceil L_c \rceil$	$\#L_c$
CYSECALERT	13 306	136	119	6	486	2
MSE	3 001	284	277	57	686	2
THREATREPORT	461	4 370	3 366	7	26 853	39
SBR	5 000	887	458	29	32 785	5
SMS	5 574	80	61	2	910	2

For security-centric analysis, we utilize three primary datasets: CYSECALERT [11], Mean Squared Error (MSE) [32], and THREATREPORT (self-labeled). The CYSECALERT and MSE datasets comprise security-related short messages extracted from X (formerly Twitter). The THREATREPORT dataset encompasses security-related content aggregated from news outlets and security feeds. While the former two datasets are representative for CERT data aggregations in crisis, the third represents data of the daily work of CERTs. In both areas the volume of information increased tremendously in recent years, while understaffing remained on a high level [29]. For product-specific analysis, the SBR dataset contains security-related messages from issue trackers spanning five distinct products [398]. Both use-cases To establish a baseline for general text classification, we incorporate the UCI *SMS Spam Collection* [399], which features characteristics common to security domain texts, including abbreviations, non-standard nomenclature, and spam content.

The labeling of THREATREPORT was done by two researchers in the field of information security. After the first independent labeling of 10 data points, both researchers discussed their labeling process and aligned differences. Afterward, both researchers continued independent on a half dataset each.

13.3.2 Operational Requirements for Automated CERT Information Processing

The exponential growth in security-related data requires CERTs to conduct extensive manual analysis daily, leading to significant operational strain [27, 28]. This sustained cognitive load frequently results in professional burnout and potential workforce attrition. The automation of routine analytical tasks presents a critical

"CyberRange : The Open-Source AWS Cyber Range [...]"

(a) Example text for the CySecAlert dataset (use-case I).

"SMBs need to take immediate action on #microsoft #exchange #vulnerabilities [URL] [...]"

(b) Example text for the Mean Squared Error dataset (use-case I).

"New CacheWarp AMD CPU attack lets hackers gain root in Linux VMs- November 14, 2023- 03:34 PM-2 A new software-based fault injection attack, CacheWarp, can let threat actors hack into AMD SEV-protected [...]"

(c) Example text for the ThreatReport dataset (use-case I).

"`SYSCS_UTIL.SYSCS_COMPRESS_TABLE` *should create statistics if they do not exist There must be an entry in the* `SYSSTATISTICS` *table in order for the cardinality statistics in* `SYSSTATISTICS` *to be created with* `SYSCS_UTIL.SYSCS_COMPRESS_TABLE` `SYSCS_UTIL.SYSCS_COMPRESS_TABLE` *should create statistics if they don't exist. [...]*"

(d) Example text for the security bug report dataset (use-case II).

"Auction round 4. The highest bid is now £54. Next maximum bid is £71. To bid, send BIDS e. g. 10 (to bid £10) to 83383. Good luck"

(e) Example text for the SMS dataset (use-case III).

Listing 13.1 Example texts from the different evaluation datasets (CySecAlert, MSE, ThreatReport, SBR, and SMS).

opportunity for operational improvement, particularly in the identification of duplicate and related information within incoming data streams. Research indicates that security personnel estimate "half of their tasks to all of their tasks could be automated today" [27]. Drawing from multiple empirical studies [12, 27, 28, 184], we establish the following core requirements for an effective CERT clustering system.

R1 **Reducing information overload for CERTs.**: The clustering system must demonstrably reduce the volume of information requiring manual review through effective cluster consolidation. Cluster homogeneity must be maximized through rigorous outlier management. The presence of misclassified data points would significantly compromise cluster integrity and negate the intended benefits of information reduction. Therefore, the system must prioritize classification accuracy over cluster completeness.

R2 **Unburden CERTs.**: The proposed algorithms must operate with minimal configuration requirements, eliminating the need for continuous model adjustment with emerging vulnerabilities or technologies. Model fine-tuning represents a significant operational overhead. While potentially more engaging than routine document review, such tasks divert resources from core responsibilities: threat identification, analysis, and stakeholder communication.

R3 **Retention of data.**: All alerts must remain accessible, regardless of their cluster assignability. The system must preserve outliers during analysis rather than forcing them into inappropriate clusters. Otherwise, important information might be either missed, due to being assigned to an outlier cluster or simply confuses personnel if found in wrong clusters. Either case would diminish the benefits of clustering the data due to lost trust in the system. This requirement aligns with information overload reduction by enabling a discrete outlier cluster for manual review, rather than discarding or misclassifying these data points.

R4 **Runtime performance.**: While not the primary optimization target, the system must complete clustering operations on new data and present results within an operationally acceptable timeframe, defined here as several minutes. The system might be run on-demand by CERT personnel in preparation of the daily inbound information review.

13.3.3 Embeddings, Clustering, and Evaluation

For the embedding process, we evaluate a diverse range of locally deployable LLMs. The selected models span from lightweight architectures with 33.4M parameters (all-MiniLM-L12-v1) to large-scale models with 7.85B parameters (NV-Embed-v2) [2, 45, 404–410]. We perform all embeddings using *sentence-transformers* [45] with the models enumerated in Table 13.2, utilizing computing infrastructure equipped with either NVidia A100 or NVidia H100 GPUs. Initial attempts to process the embeddings on an Apple M4 system with 24 GB memory proved insufficient due to memory constraints.

Table 13.2 Overview of the LLMs used in combination with SBERT (sorted alphabetically)

Huggingface Model ID ↓	Params
Alibaba-NLP/gte-base-en-v1.5	137M
jxm/cde-small-v1	281M
markusbayer/cysecbert	110M
meta-llama/Llama-3.2-1B	1.24B
meta-llama/Llama-3.2-3B	3.21B
mistralai/Mistral-7B-v0.1	7.24B
mistralai/Mistral-7B-v0.3	7.25B
NovaSearch/stella_en_15b_v5	1.54B
NovaSearch/stella_en_400M_v5	435M
nvidia/NV-Embed-v2	7.85B
sentence-transformers/all-MiniLM-L12-v1	33.4M
sentence-transformers/all-mpnet-base-v2	109M
thenlper/gte-large	335M
sentence-transformers/gtr-t5-xxl	4.86B

Our clustering methodology incorporates algorithms from four major categories: *partition-based clustering*, *density-based clustering*, *hierarchical clustering*, and *deep clustering* [378], selected based on their prevalence, operational characteristics (*e.g.*, arbitrary cluster shape detection, parameter complexity), and MTEB ranking[2]. We prioritize algorithmic simplicity to assess performance under minimal parameter optimization. This approach aligns with 2̊, enabling CERTs to focus on core responsibilities without extensive hyperparameter tuning. Table 13.3 provides a comprehensive overview of the selected clustering algorithms. To evaluate the requirements 2̊ and 4̊ regarding operational burden and runtime performance, we conduct clustering experiments on an Apple M4 system with 24 GB memory. This

[2] https://huggingface.co/spaces/mteb/leaderboard

represents a worst-case scenario using consumer hardware, below typical CERT infrastructure capabilities.

Table 13.3 Overview of used cluster algorithms, a short explanation, and the used parameters. If no parameter setting is displayed, we use the defaults selected by ClustPy [19]

Algorithm	Description	Parameters
	Partition-based Clustering	
K-Means [378]	Divides data into k clusters by minimizing within-cluster variance. Uses centroid-based approach with spherical cluster assumptions.	n_cluster = 12
Bruteforce-K-Means	Bruteforce to find optimal $k \in \left[2, \lceil\sqrt{\lvert c\rvert}\rceil\right]$ based on resulting silhouette score, where c is the given dataset.	
SpecialK [399]	Determining the optimal number of clusters by developing a probabilistic method to assess if clusters originate from a single distribution	
SkinnyDip [401]	Noise-robust clustering algorithm designed for datasets with up to 80% noise, using Hartigan's dip test of unimodality and recursive univariate projection analysis.	
	Hierarchical Clustering	
Agglomerative Clustering [402]	Hierarchical bottom-up clustering merging closest data points, creating cluster hierarchy via dendrogram.	n_clusters = 12
	Density-based Clustering	
DBSCAN [383]	Clusters dense regions separated by low-density areas. Robust to outliers, discovers arbitrarily shaped clusters	eps = 0.5 min_samples = 2 metric = precomputed
OPTICS [403]	Advanced density-based clustering handling varying cluster densities. Creates reachability plot for comprehensive structure analysis.	min_samples = 5 metric = precomputed
	Deep Clustering	
DKM [389]	Deep clustering approach that jointly learns data representations and cluster assignments through a continuous reparametrization for k-Means, solely relying on gradient descent.	n_cluster = 12

(Continued)

Table 13.3 (Continued)

Algorithm	Description	Parameters
DDC [388]	Two-stage deep density-based image clustering framework using convolutional autoencoder and t-SNE for low dimensionality and density-based clustering to recognize clusters.	
DipEncoder [386]	Coupling the Hartigan's unsupervised Dip-test with an autoencoder to obtain cluster embeddings.	n_cluster = 12
N2D [385]	Learns autoencoded embedding, uses UMAP for manifold learning, and applies shallow clustering algorithms.	n_cluster = 12
DeepECT [384]	Builds cluster tree in embedding space, which allows selecting the number of clusters afterwards.	
DEC [390]	Neural network clustering with iterative representation and cluster refinement.	n_cluster = 12
DipDECK [387]	Advanced deep clustering with disentangled representation approach.	

The evaluation framework employs exclusively external metrics for two fundamental reasons. First, internal metrics introduce inherent bias when comparing diverse clustering algorithms, as certain algorithms optimize specific internal criteria (*e.g.*, K-Means for silhouette coefficient). Second, intrinsic evaluation metrics assess cluster morphology rather than semantic accuracy relative to ground truth. For CERT applications, semantic cohesion within clusters is paramount to prevent analytical confusion and redundant verification. We adopt the external metrics proposed by Rosenberg and Hirschberg [396]: *homogeneity*, *completeness*, and *V-measure*, with primary emphasis on *homogeneity*. Homogeneity $h \in [0, 1] \subseteq \mathbb{R}$ quantifies intra-cluster uniformity, achieving its maximum of 1 when clusters perfectly align with ground truth [396]. While completeness $c \in [0, 1] \subseteq \mathbb{R}$ measures object distribution across clusters, our focus on precision renders this metric secondary. V-measure $V_\beta = \frac{(1+\beta)hc}{(\beta * h)+c} \subseteq \mathbb{R}$ combines these metrics through their harmonic mean, reaching 1 for optimal clustering. We set $\beta = 0$, effectively reducing V-measure to homogeneity, prioritizing semantic consistency for CERT operations.

Our comprehensive evaluation encompasses 14 embedding models, 14 clustering algorithms across 5 datasets, measuring 8 distinct metrics while prioritizing *homogeneity*. Additional captured metrics include: *completeness*, *V-measure*, *silhouette coefficient*, *Adjusted Rand Index*, *Calinski-Harabasz index*, *Davies-Bouldin index*, and *runtime performance*. All metrics represent averages across 5 consecutive iterations, totaling 4900 distinct experimental configurations.

13.4 Evaluation

Our evaluation methodology comprises three primary components. First, we analyze clustering performance across three distinct dataset categories: (i) CTI datasets, (ii) SBRs, and (iii) general short message data (UCI's SMS dataset). Second, we assess computational efficiency through runtime analysis of both embedding generation and clustering operations. Finally, we correlate these findings with our research questions and established requirements in the subsequent section.

13.4.1 Clustering Performance

Our analysis encompasses 196 distinct cluster-embedding combinations for each dataset, with results averaged across 5 consecutive executions. Clustering performance exhibits significant variation correlated with ground truth cluster cardinality and dataset dimensionality. Table 13.4 presents the five highest-performing cluster-embedding combinations per dataset.

The CySecAlert dataset demonstrates superior performance with partition-based clustering methodologies. Model capacity shows limited correlation with performance, as evidenced by comparable results between *NV-Embed-v2* (7.85B parameters) and *stella_en_400M_v5* (435M parameters). While homogeneity metrics achieve a maximum of 0.61, additional metrics indicate significant cluster overlap. Completeness remains below 0.11, yielding a maximum V-measure of 0.19. Silhouette coefficient, Adjusted Rand Index, and Calinski-Harabasz index collectively indicate suboptimal cluster separation. The optimal configuration generates 12 clusters, achieving a 99.91% reduction in dataset cardinality.

For the MSE dataset, density-based clustering, specifically OPTICS, achieves optimal performance. Homogeneity reaches 0.49, with superior performance from models below 1.24B parameters (Llama-3.2-1B) to 33M parameters. Auxiliary metrics suggest cluster overlap challenges, while achieving 94.80% input reduction. The ThreatReport dataset exhibits optimal performance with deep and density-based clustering, achieving maximum homogeneity of 0.34. These results indicate insufficient performance for production deployment, necessitating enhanced fine-

Table 13.4 Evaluation of the datasets sorted by homogeneity (H) with ranking (#). It shows the used evaluation combination: clustering algorithm (Algorithm), embedding model (Model), and metrics. The columns denote homogeneity (H), completeness (C), V-measure (V-M), Silhouette coefficient (Sil), Adjusted Rand Index (ARI), Calinski-Harabasz index (CH), Davies-Bouldin index (DB), runtime in seconds (t [s]), and number of predicted clusters (#C), respectively. Results are sorted in descending order by homogeneity (↓). After the evaluation we evaluated the THREATREPORT dataset with the parameters *n_clusters = 60*, denoted as THREATREPORT-60

	Algorithm	Model	H ↓	C	V–M	Sil	ARI	CH	DB	t [s]	#C
CYSECALERT (GT: $\|c\| = 13306$, #C = 2)—Use-case I											
1	K-Means	NV-Embed-v2	0.61	0.10	0.17	0.01	0.04	177.35	5.27	2.75	12.0
2	BruteForceK-Means	NV-Embed-v2	0.59	0.11	0.19	0.01	0.04	240.24	5.42	1.65	8.0
3	K-Means	stella_en_400M_v5	0.55	0.09	0.15	0.02	0.04	205.96	4.82	0.84	12.0
4	K-Means	stella_en_1.5b_v5	0.54	0.09	0.15	–0.01	0.03	212.45	4.85	0.88	12.0
5	BruteForceK-Means	stella_en_400M_v5	0.54	0.10	0.17	0.02	0.05	280.10	4.99	0.61	8.0
MSE (GT: $\|c\| = 3001$, #C = 2)—Use-case I											
1	OPTICS	llama-3.2-1b	0.49	0.10	0.17	0.26	0.03	16.87	1.43	3.99	156.0
2	OPTICS	all-minilm-l12-v2	0.48	0.10	0.17	0.24	0.03	17.20	1.41	4.04	150.0
3	OPTICS	all-mpnet-base-v2	0.45	0.10	0.16	0.21	0.03	16.81	1.40	3.99	145.0
4	OPTICS	NV-Embed-v2	0.45	0.10	0.16	0.17	0.02	13.74	1.53	4.01	161.0
5	OPTICS	gte-base-en-v15	0.45	0.11	0.17	0.20	0.03	16.53	1.39	3.96	135.0

(Continued)

Table 13.4 (Continued)

	Algorithm	Model	H ↓	C	V–M	Sil	ARI	CH	DB	t [s]	#C
		ThreatReport (GT: $\lvert c \rvert = 461$, #C = 39)—Use-case I									
1	DDC	gte-base-en-v15	0.34	0.30	0.32	0.02	0.03	6.90	2.26	2.90	30.4
2	DDC	all-minilm-l12-v2	0.33	0.29	0.31	–0.01	0.02	6.88	2.08	2.54	34.0
3	OPTICS	gte-large	0.32	0.34	0.33	0.08	0.05	9.28	2.16	0.10	23.0
4	OPTICS	stella_en_1.5b_v5	0.31	0.36	0.34	0.01	0.06	7.67	2.01	0.14	25.0
5	OPTICS	gte-base-en-v15	0.31	0.35	0.33	0.08	0.06	7.50	2.07	0.12	25.0
		ThreatReport–60 (GT: $\lvert c \rvert = 461$, #C = 39)—Use-case I									
1	AgglomerativeClustering	gte-base-en-v15	0.51	0.33	0.40	0.27	0.02	9.26	1.65	0.03	60
2	AgglomerativeClustering	all-mpnet-base-v2	0.50	0.33	0.40	0.27	0.02	10.59	1.55	0.03	60
3	AgglomerativeClustering	stella_en_400M_v5	0.50	0.33	0.40	0.28	0.02	10.91	1.46	0.04	60
4	N2D	gte-base-en-v15	0.50	0.32	0.39	0.20	0.01	7.38	1.99	2.92	60
5	AgglomerativeClustering	gtr-t5-xxl	0.50	0.32	0.39	0.26	0.02	9.25	1.63	0.02	60
		Security bug report (GT: $\lvert c \rvert = 5000$, #C = 5)—Use-case II									
1	DEC	stella_en_1.5b_v5	0.88	0.63	0.74	0.02	0.64	91.20	4.94	39.66	12.0
2	DKM	stella_en_1.5b_v5	0.87	0.62	0.72	–0.07	0.61	87.08	5.59	33.41	12.0
3	DeepECT	stella_en_1.5b_v5	0.86	0.47	0.61	–0.00	0.35	62.77	5.06	68.39	20.0
4	DEC	stella_en_400M_v5	0.86	0.60	0.71	0.02	0.60	96.50	6.44	39.52	12.0
5	DEC	gte-base-en-v15	0.84	0.58	0.68	0.04	0.57	93.44	4.97	34.88	12.0
		SMS (GT: $\lvert c \rvert = 5574$, #C = 2)—Use-case III									
1	K-Means	NV-Embed-v2	0.89	0.15	0.25	0.02	0.06	78.78	4.73	2.42	12.0
2	DKM	NV-Embed-v2	0.88	0.46	0.60	0.01	0.76	38.83	5.90	64.98	12.0
3	BruteForceK-Means	NV-Embed-v2	0.88	0.18	0.29	0.03	0.09	103.05	4.95	1.22	8.0
4	DDC	NV-Embed-v2	0.88	0.24	0.37	0.01	0.27	77.20	4.76	53.21	8.8
5	DipEncoder	NV-Embed-v2	0.88	0.14	0.24	0.02	0.05	71.92	5.44	159.85	12.0

tuning procedures. Supplementary metrics corroborate suboptimal clustering performance. Despite achieving 93.41% dimensional reduction, the clustering quality remains inadequate for operational deployment.

Deep clustering demonstrates superior performance on the SBR dataset, with *stella* models (*cf.* Table 13.2) achieving homogeneity exceeding 0.86. This performance suggests that large-scale models like *NV-Embed-v2* are not prerequisite for optimal embedding generation. The configuration achieves 99.76% dimensional reduction while maintaining high homogeneity. However, auxiliary metrics indicate persistent cluster overlap challenges, with silhouette coefficients approximating 0.

For the general-purpose SMS dataset, partition-based and deep clustering algorithms paired with *NV-Embed-v2* achieve optimal performance. K-Means clustering achieves 99.78% input reduction with exceptional homogeneity exceeding 0.88 across top-five configurations. However, secondary metrics continue to indicate cluster overlap challenges (Table 13.5).

Table 13.5 Runtime results using the mean value of a total of 5 iterations for all datasets

	K-Means	AgglomerativeClustering	OPTICS	BruteForceK-Means	SkinnyDip	DDC	DEC	DeepECT	DipDECK	DipEncoder	DKM	N2D	DBSCAN	SpecialK
CySecAlert (GT: $\lvert c \rvert = 13\,306$, #C = 2) — Use-case I														
Min	0.54	8.08	302.67	0.40	0.06	76.83	92.82	180.59	56.18	219.19	78.87	73.20	-	-
Max	4.30	80.28	313.88	2.94	0.40	131.88	184.92	377.33	186.90	344.12	174.94	133.67	-	-
Mean	1.51	37.04	307.16	1.09	0.18	95.09	123.81	236.20	113.20	265.79	109.43	92.34	-	-
MSE (GT: $\lvert c \rvert = 3\,001$, #C = 2) — Use-case I														
Min	0.11	0.45	3.96	0.07	1.05	14.40	17.65	38.47	27.49	55.66	15.43	15.07	0.07	-
Max	1.26	4.02	4.25	1.05	13.16	28.29	41.53	111.69	58.35	85.04	37.79	28.87	0.10	-
Mean	0.45	1.80	4.03	0.32	5.42	19.39	26.34	55.90	35.69	67.15	23.04	19.63	0.08	-
ThreatReport (GT: $\lvert c \rvert = 461$, #C = 39) — Use-case I														
Min	0.04	0.01	0.10	0.04	0.09	2.54	3.42	3.68	6.90	6.02	2.53	2.73	0.00	-
Max	0.32	0.15	0.20	0.21	1.31	4.78	8.15	15.24	14.11	12.07	6.50	5.94	0.00	-
Mean	0.11	0.06	0.13	0.08	0.48	3.37	4.88	8.57	9.25	8.38	4.02	3.59	0.00	-
Security bug report (GT: $\lvert c \rvert = 5\,000$, #C = 5) — Use-case II														
Min	0.15	1.13	16.38	0.12	-	25.29	31.93	61.09	21.89	91.79	27.46	25.26	-	-
Max	0.74	2.99	17.33	1.17	-	30.70	39.66	88.85	48.51	105.60	33.83	29.55	-	-
Mean	0.37	2.36	16.99	0.34	-	28.24	36.80	70.54	36.37	95.85	31.33	28.09	-	-
SMS (GT: $\lvert c \rvert = 5\,574$, #C = 2) — Use-case III														
Min	0.25	1.38	140.46	0.16	0.03	29.14	35.31	76.86	42.70	107.83	29.81	27.18	1.28	4.20
Max	2.49	13.43	176.86	1.59	1.01	55.41	78.64	176.91	94.06	171.19	73.11	58.07	1.44	21.56
Mean	0.92	6.70	149.56	0.69	0.34	39.96	54.16	112.33	64.32	128.59	46.51	40.02	1.39	13.83

Figure 13.1 depicts the best and worst performing security related datasets. The rest is depicted in Appendix B.

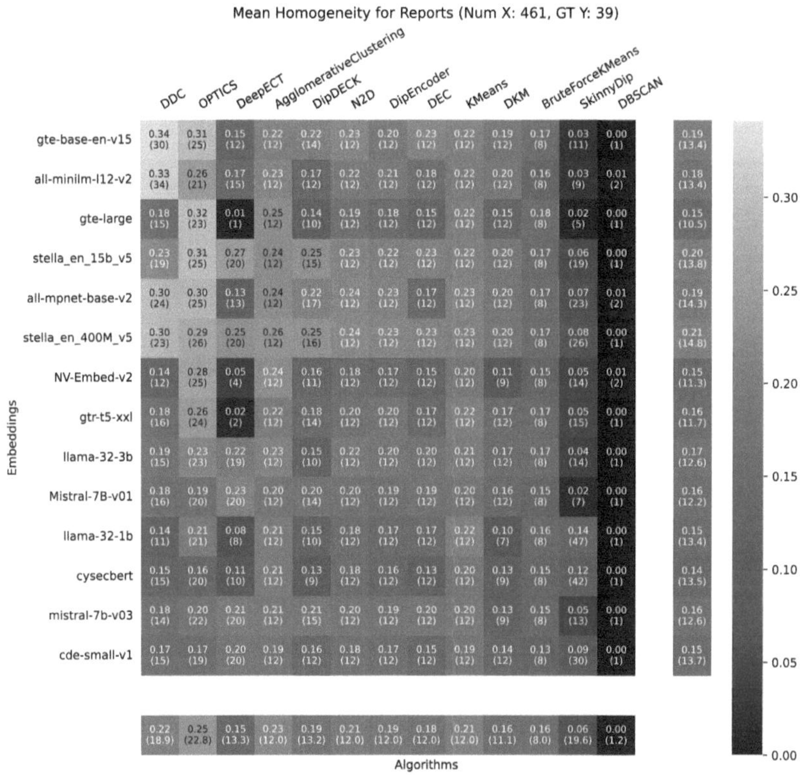

(a) Mean homogeneity of the THREATREPORT dataset.

Fig. 13.1 Result of the best and worst performing security related datasets with regard to the mean homogeneity over 5 consecutive runs. Columns show the clustering algorithms and rows the used embeddings. The separated column and row depict the mean over each column and row, respectively. The models and algorithms are sorted by the rows/columns sum (descending), such that the top left shows the highest result, while the bottom right shows the worst.

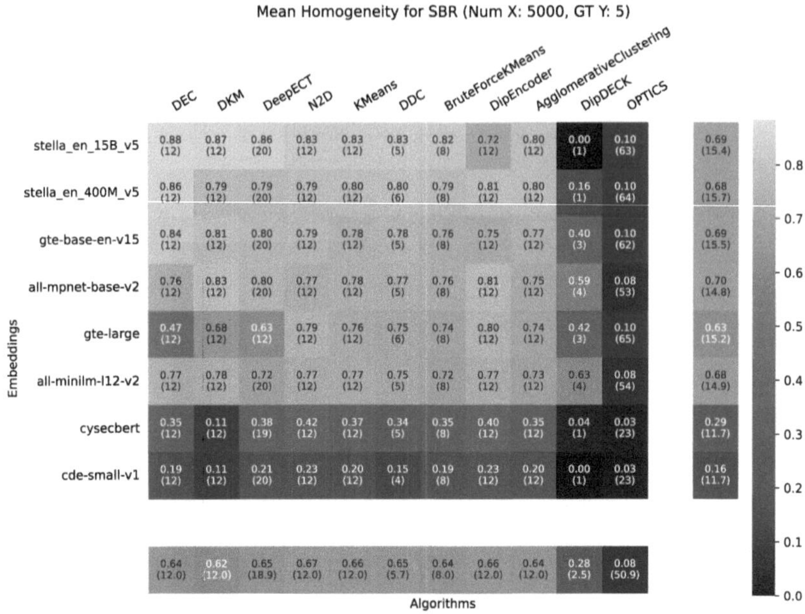

(b) Mean homogeneity of the SBR dataset.

Fig. 13.1 (Continued)

13.4.2 Runtime Performance

Runtime analysis encompasses both embedding generation and clustering operations, measured in seconds. Both phases demonstrate acceptable computational efficiency (*cf.* Tables 13.4 and 13.6). Embedding generation peaks at 287 s for the CySecAlert dataset, while requiring only 80 s for the SMS dataset. The CySecAlert dataset exhibits minimum embedding time of 14.31 s with mean execution time of 84.15 s. When combined with clustering operations, DipEncoder requires maximum 344.12 s, while K-Means achieves optimal performance in 4.3 s, yielding total pipeline execution under 300 s. Comparable performance characteristics are observed across optimal cluster-embedding combinations: MSE achieves maximal runtime of 235 s (embed + OPTICS), ThreatReport requires 220 s (embed + DDC), SBR completes in 300 s (embed + DeepECT), and SMS processing concludes in 155 s (embed + DKM).

Table 13.6 Tabular display of runtime statistics of embedding the different datasets

Dataset	Min	Max	Mean	Median	Std
CYSECALERT	14.31	287.40	84.15	48.58	93.60
MSE	5.01	230.32	51.72	17.86	68.36
THREATREPORT	4.05	215.21	55.26	24.90	75.88
SBR	11.18	210.71	68.47	40.83	63.69
SMS	7.28	80.53	32.89	23.86	27.44

13.5 Discussion, Limitations, and Future Work

This research evaluates the efficacy of clustering methods in mitigating information overload for CERT personnel. While our evaluation demonstrates promising results, several aspects warrant detailed discussion and highlight opportunities for future research.

13.5.1 Discussion

Our research question "Which cluster algorithm and embedding combination is suitable to reduce CERT personnel's information overload?" can be answered with qualified success. The evaluation demonstrates that no single combination universally excels across all CTI datasets, though several configurations show promising results. For immediate operational deployment, deep clustering combined with *stella* models achieves optimal performance on SBR data (homogeneity .84), while partition-based clustering with *NV-Embed-v2* performs well on general advisory data (CYSECALERT, homogeneity 0.61). However, the core CERT focus on threat reports (THREATREPORT) requires careful parameter tuning, achieving homogeneity of 0.5 only with adjusted cluster counts.

Information overload represents a primary factor in CERT personnel burnout and attrition [27]. To address this challenge, we established clustering requirements through comprehensive analysis of current research [12, 27, 184, 375]. Our methodology incorporated both established and novel clustering approaches, leveraging diverse LLMs for document embedding (*cf.* Tables 13.2 and 13.3). The evaluation framework prioritized homogeneity as an external metric [396], analyzing three distinct CTI datasets [11, 32], including our novel THREATREPORT dataset (**C1**), alongside SBR data [398, 411, 412] and the UCI SMS corpus [399].

Our results indicate that achieving homogeneous clustering for CTI advisories and reports presents significant challenges. While CySecAlert and MSE datasets achieve homogeneity scores of 0.61 and 0.49 respectively, these metrics do not fully address primary CERT operational requirements. The ThreatReport dataset achieves only 0.34 homogeneity with default parameters (*cf.* Table 13.3), resulting in mixed clusters and complex structures (indicated by near-zero silhouette scores). Subsequent analysis with adjusted parameters (n_clusters = 60 versus default 12) for the ThreatReport dataset yields substantially improved results, surpassing 0.5 homogeneity and approaching CySecAlert and MSE performance (*cf.* ThreatReport-60 in Table 13.4). The SBR dataset demonstrates exceptional potential with homogeneity exceeding 0.84, though cluster structure remains complex (silhouette scores approximating 0). Deep clustering methodologies effectively address these structural challenges, producing promising results. These findings suggest clustering offers domain-specific utility (*e.g.*, SBR), while CERT applications may require algorithmic fine-tuning or careful dataset curation. The methodology demonstrates strong generalization, evidenced by SMS dataset homogeneity exceeding 0.85.

All evaluated algorithms achieve data reduction exceeding 90%, directly addressing requirement 1̊ for information overload mitigation. Even with adjusted parameters, the ThreatReport dataset maintains reduction rates above 86.98%, validating clustering efficacy in CERT operations. The simplified configuration aligns with requirement 2̊, reducing operational burden while preserving original data points 3̊.

Runtime performance exceeds expectations, with complete processing requiring approximately 5 minutes. This efficiency could potentially save a single CERT over 3750 hours annually[3]. This performance satisfies requirement 4̊ while preserving operational capacity for core CERT infrastructure security responsibilities.

13.5.2 Limitations and Future Work

Our methodology exhibits several key limitations. Parameter optimization was intentionally omitted to align with requirement 2̊, though results suggest its necessity for heterogeneous CTI report processing. Our evaluation methodology is constrained by the limited number of consecutive runs (5), hardware restrictions (Apple

[3] Given the process is done once a day by a single person for the whole CERT, every workday in a year. Kaufhold et al. [28] stated a daily workload of up to 2 hours, which is reduced to 5 minutes which can be run asynchronously plus at most 30 minutes information sourcing: 250×1.5 hours = 3750 hours.

M4, 24 GB memory), and consistently poor cluster separation indicated by silhouette scores approximating 0.

Dataset limitations include potential bias in the manually labeled THREATREPORT dataset and significant performance sensitivity to cluster count parameters (12 versus 60 clusters). While achieving high dimensional reduction (>90%), the long-term impact of false negatives on CERT operations remains unevaluated. Additionally, we excluded domain-specific features(*e.g.*, CVE identifiers, OWASP classifications) and foundational models (*e.g.*, Claude, DeepSeek, GPT) from our analysis.

Future work should address these limitations through expanded stability analysis, alternative distance metrics, systematic parameter sensitivity evaluation, and integration of domain-specific features while maintaining operational simplicity. The exploration of multilingual capabilities and assessment of false negative impact on operational efficiency present additional research opportunities.

13.6 Conclusion

This work investigated the application of clustering algorithms for reducing information overload in CERT operations. Through comprehensive evaluation of 196 cluster-embedding combinations across five datasets, we demonstrate that clustering can effectively reduce information processing requirements by over 90% while maintaining semantic coherence. However, optimal performance requires careful selection of clustering approaches based on specific data characteristics.

Deep clustering combined with *stella* models demonstrates superior performance for structured security data (SBR, homogeneity 0.84), while partition-based clustering with *NV-Embed-v2* excels for general advisory content (CYSECALERT, homogeneity 0.61). The more complex THREATREPORT dataset requires parameter adjustment to achieve acceptable performance (homogeneity 0.5), highlighting the need for domain-specific tuning. Runtime performance remains consistently efficient, with complete processing requiring approximately five minutes, potentially saving CERTs over 3750 hours annually.

While our evaluation demonstrates clustering's potential for information overload reduction, several challenges remain. Future work should address cluster separation optimization, systematic parameter tuning, and integration of domain-specific features while maintaining operational simplicity. Despite these limitations, our findings suggest that clustering approaches, when properly configured, can significantly enhance CERT operational efficiency without compromising analytical integrity.

Bibliography

1. **Kuehn, P.,** Büumler, J., Kaufhold, M.-A., Wendelborn, M., Reuter, C., "The Notion of Relevance in Cybersecurity: A Categorization of Security Tools and Deduction of Relevance Notions," in Workshop-Proceedings Mensch und Computer, Darmstadt: Gesellschaft für Informatik, 2022. https://doi.org/10.18420/muc2022-mci-ws01-220.
2. Bayer, M., **Kuehn, P.,** Shanehsaz, R., Reuter, C., "CySecBERT: A Domain- Adapted Language Model for the Cybersecurity Domain," *ACM Transactions on Privacy and Security*, Mar. 15, 2024. https://doi.org/10.1145/3652594.
3. **Kuehn, P.,** Nadermahmoodi, D., Bayer, M., Reuter, C., "Bandit on the Hunt: Dynamic Crawling for Cyber Threat Intelligence," *arXiv*, Apr. 25, 2025. https://doi.org/10.48550/arXiv.2504.18375.
4. **Kuehn, P.,** Wittorf, K., Reuter, C., "Navigating the Shadows: Manual and Semi-Automated Evaluation of the Dark Web for Cyber Threat Intelligence," *IEEE Access*, vol. 12, pp. 118 903–118 922, 2024. https://doi.org/10.1109/ACCESS.2024.3448247.
5. **Kuehn, P.,** Bayer, M., Wendelborn, M., Reuter, C., "OVANA: An Approach to Analyze and Improve the Information Quality of Vulnerability Databases," in *ARES '21: Proceedings of the 16th International Conference on Availability, Reliability and Security*, ACM, 2021, p. 11. https://doi.org/10.1145/3465481.3465744.
6. **Kuehn, P.,** Relke, D. N., Reuter, C., "Common vulnerability scoring system prediction based on open source intelligence information sources," *Computers & Security*, 2023. https://doi.org/10.1016/j.cose.2023.103286.
7. **Kuehn, P.,** Bayer, M., Frey, T., Kerk, M., Reuter, C., "Reducing Information Overload: Because Even Security Experts Need to Blink," in *Availability, Reliability and Security—ARES 2025 Workshops*, Ghent, Belgium: Springer Nature, 2025.
8. **Kuehn, P.,** Riebe, T., Apelt, L., Jansen, M., Reuter, C., "Sharing of Cyber Threat Intelligence between States," *S+F Sicherheit und Frieden / Peace and Security*, vol. 38, pp. 22–28, 2020. https://doi.org/10.5771/0175-274X-2020-1-22.
9. Egert, R., Gerber, N., Haunschild, J., **Kuehn, P.,** Zimmermann, V., "Towards Resilient Critical Infrastructures—Motivating Users to Contribute to Smart Grid Resilience," *i-com—Journal of Interactive Media*, vol. 20, pp. 161–175, 2021. https://doi.org/10.1515/icom-2021-0021.
10. Kaufhold, M.-A., Fromm, J., Riebe, T., Mirbabaie, M., **Kuehn, P.,** Basyurt, A. S., Bayer, M., Stöttinger, M., Eyilmez, K., Möller, R., Fuchß, C., Stieglitz, S., Reuter,

P. Kühn, *Proactive Cyber Threat Intelligence*, Technology, Peace and Security | Technologie, Frieden und Sicherheit, https://doi.org/10.1007/978-3-658-43951-4

C., "CYWARN: Strategy and Technology Development for Cross-Platform Cyber Situational Awareness and Actor-Specific Cyber Threat Communication," in *Workshop-Proceedings Mensch und Computer*, Bonn: Gesellschaft für Informatik, 2021. https://doi.org/10.18420/muc2021-mci-ws08-263.
11. Riebe, T., Wirth, T., Bayer, M., **Kuehn, P.,** Kaufhold, M.-A., Knauthe, V., Guthe, S., Reuter, C., "CySecAlert: An Alert Generation System for Cyber Security Events Using Open Source Intelligence Data," in *Information and Communications Security (ICICS)*, 2021, pp. 429–446. https://doi.org/10.1007/978-3-030-86890-1_24.
12. Basyurt, A. S., Fromm, J., **Kuehn, P.,** Kaufhold, M.-A., Mirabaie, M., "Help Wanted—Challenges in Data Collection, Analysis and Communication of Cyber Threats in Security Operation Centers," in *Proceedings of the International Conference on Wirtschaftsinformatik (WI)*, Nürnberg, 2022. [Online]. Available: https://aisel.aisnet.org/wi2022/it_for_development/it_for_development/20/.
13. Riebe, T., **Kuehn, P.,** Imperatori, P., Reuter, C., "U.S. Security Policy: The Dual-Use Regulation of Cryptography and its Effects on Surveillance," *European Journal for Security Research*, 2022.https://doi.org/10.1007/s41125-022-00080-0.
14. Reinhold, T., **Kuehn, P.,** Günther, D., Schneider, T., Reuter, C., "Ex- TRUST: Reducing Exploit Stockpiles With a Privacy-Preserving Depletion Systems for Inter-State Relationships," *IEEE Transactions on Technology and Society*, vol. 4, pp. 158–170, 2023. https://doi.org/10.1109/TTS.2023.3280356.
15. **Kuehn, P.,** Schmidt, M., Bayer, M., Reuter, C., "ThreatCrawl: A BERTbased Focused Crawler for the Cybersecurity Domain." (Apr. 26, 2023).
16. Reuter, C., Franken, J., Reinhold, T., **Kuehn, P.,** Kaufhold, M.-A., Riebe, T., Hartwig, K., Biselli, T., Schmid, S., Guntrum, L., Haesler, S., "Informatik für den Frieden: Perspektive von PEASEC zu 40 Jahren FIfF," *FIfF-Kommunikation*, 2024. [Online]. Available: https://peasec.de/paper/2024/2024_Reuteretal_InformatikFuerFrieden_fiff.pdf.
17. Dempsey, M. E. "Joint Intelligence." (Oct. 22, 2013), [Online]. Available: https://web.archive.org/web/20160613010839/http://www.dtic.mil/doctrine/new_pubs/jp2_0.pdf.
18. Evtimov, R., Falli, M., Maiwald, A., "Bidirectional Encoder Representations from Transformers (BERT)," Anti Social Online Behaviour Detection with BERT. (Feb. 7, 2020), [Online]. Available: https://humboldtwi.github.io/blog/research/information_systems_1920/bert_blog_post/.
19. Leiber, C., Miklautz, L., Plant, C., Böhm, C., "Benchmarking Deep Clustering Algorithms With ClustPy," in *2023 IEEE International Conference on Data Mining Workshops (ICDMW)*, Dec. 2023, pp. 625–632. https://doi.org/10.1109/ICDMW60847.2023.00087.
20. IBM, "Average cost of a data breach in the United States from 2006 to 2024," Statista. (Jul. 2024), [Online]. Available: https://www.statista.com/statistics/273575/us-average-cost-incurred-by-a-data-breach/.
21. Statista, "Cybersecurity -Worldwide," Statista. (2025), [Online]. Available: https://www.statista.com/outlook/tmo/cybersecurity/worldwide.
22. Reuters, "Cyber attack compromised Indonesia data centre, ransom sought," *ReutersCybersecurity*, Jun. 24, 2024. [Online]. Available: https://www.reuters.com/technology/cybersecurity/cyber-attack-compromised-indonesia-data-centre-ransom-sought-reports-antara-2024-06-24/.
23. Dobberstein, L. "Hackers of Indonesian government apologize and give key," The Register. (Jul. 4, 2024), [Online]. Available: https://www.theregister.com/2024/07/04/hackers_of_indonesian_government_apologize/.

24. ISO, "Threat intelligence and why it matters for cybersecurity," ISO. (2025), [Online]. Available: https://www.iso.org/information-security/threat-intelligence.
25. A. Dehghantanha, M. Conti, and T. Dargahi, Eds., *Cyber Threat Intelligence*. Cham: Springer International Publishing, 2018, vol. 70, p. 628, 623 pp. https://doi.org/10.1007/978-3-319-73951-9.
26. McMillan, R. "Definition: Threat Intelligence," Gartner. (May 16, 2013), [Online]. Available: https://www.gartner.com/en/documents/2487216.
27. Hinchy, E., "Voice of the SOC Analyst," Tines, 2022, p. 39. [Online]. Available: https://www.tines.com/reports/voice-of-the-soc-analyst/.
28. Kaufhold, M.-A., Riebe, T., Bayer, M., Reuter, C., "'We Do Not Have the Capacity to Monitor All Media': A Design Case Study on Cyber Situational Awareness in Computer Emergency Response Teams," in *CHI'24*, New York, NY, USA: Association for Computing Machinery, May 11, 2024, pp. 1–16. https://doi.org/10.1145/3613904.3642368.
29. ISACA, "Companies' cybersecurity staffing worldwide in 2023 versus 2024," Statista. (Oct. 2024), [Online]. Available: https://www.statista.com/statistics/1322073/cybersecurity-staffing-worldwide/.
30. Sauerwein, C., Fischer, D., Rubsamen, M., Rosenberger, G., Stelzer, D., Breu, R., "From Threat Data to Actionable Intelligence: An Exploratory Analysis of the Intelligence Cycle Implementation in Cyber Threat Intelligence Sharing Platforms," in *Proceedings of the 16th International Conference on Availability, Reliability and Security*, New York, NY, USA: Association for Computing Machinery, Aug. 17, 2021, pp. 1–9. https://doi.org/10.1145/3465481.3470048.
31. Oosthoek, K., Doerr, C., "Cyber Threat Intelligence: A Product Without a Process?" *International Journal of Intelligence and CounterIntelligence*, vol. 34, pp. 300–315, Apr. 3, 2021. https://doi.org/10.1080/08850607.2020.1780062.
32. Bayer, M., Frey, T., Reuter, C., "Multi-level fine-tuning, data augmentation, and few-shot learning for specialized cyber threat intelligence," *Computers & Security*, vol. 134, p. 103 430, Nov. 1, 2023. https://doi.org/10.1016/j.cose.2023.103430.
33. Liao, X., Yuan, K., Wang, X., Li, Z., Xing, L., Beyah, R., "Acing the IOC game: Toward automatic discovery and analysis of open-source cyber threat intelligence," in *CSS'16*, vol. 24-28-Octo, New York, New York, USA: ACM Press, 2016, pp. 755–766. https://doi.org/10.1145/2976749.2978315.
34. Zhao, J., Yan, Q., Li, J., Shao, M., He, Z., Li, B., "TIMiner: Automatically extracting and analyzing categorized cyber threat intelligence from social data," *Computers and Security*, vol. 95, pp. 1–27, May 2017 2020. https://doi.org/10.1016/j.cose.2020.101867.
35. Debnath, B., Solaimani, M., Gulzar, M. A. G., Arora, N., Lumezanu, C., Xu, J., Zong, B., Zhang, H., Jiang, G., Khan, L., "LogLens: A Real-Time Log Analysis System," in *2018 IEEE 38th International Conference on Distributed Computing Systems (ICDCS)*, Jul. 2018, pp. 1052–1062. https://doi.org/10.1109/ICDCS.2018.00105.
36. Ranade, P., Piplai, A., Joshi, A., Finin, T., "CyBERT: Contextualized Embeddings for the Cybersecurity Domain," in *2021 IEEE International Conference on Big Data (Big Data)*, Dec. 2021, pp. 3334–3342. https://doi.org/10.1109/BigData52589.2021.9671824.
37. Agarwal, N., Yiliyasi, Y., "Information quality challenges in social media," *Proceedings of the 2010 International Conference on Information Quality, ICIQ 2010*, p. 15, 2010.

38. Anwar, A., Abusnaina, A., Chen, S., Li, F., Mohaisen, D., "Cleaning the NVD: Comprehensive Quality Assessment, Improvements, and Analyses," *IEEE Transactions on Dependable and Secure Computing*, vol. 19, pp. 4255–4269, 2022. https://doi.org/10.1109/TDSC.2021.3125270.
39. Dong, Y., Guo, W., Chen, Y., Xing, X., Zhang, Y., Wang, G., "Towards the detection of inconsistencies in public security vulnerability reports," in *Proceedings of the 28th USENIX Security Symposium*, Santa Clara, CA: USENIX Association, Aug. 2019, pp. 869–885.
40. Zhang, H., Santos, A., Freire, J., "DSDD: Domain-Specific Dataset Discovery on theWeb," in *Proceedings of the 30th ACM International Conference on Information & Knowledge Management*, New York, NY, USA: Association for Computing Machinery, Oct. 30, 2021, pp. 2527–2536. https://doi.org/10.1145/3459637.3482427.
41. Koloveas, P., Chantzios, T., Alevizopoulou, S., Skiadopoulos, S., Tryfonopoulos, C., "inTIME: A Machine Learning-Based Framework for Gathering and Leveraging Web Data to Cyber-Threat Intelligence," *Electronicsweek*, vol. 10, p. 818, 2021. https://doi.org/10.3390/electronics10070818.
42. Allen, L., O'Connell, A., Kiermer, V., "How can we ensure visibility and diversity in research contributions? How the Contributor Role Taxonomy (CRediT) is helping the shift from authorship to contributorship," *Learned Publishing*, vol. 32, pp. 71–74, Jan. 1, 2019. https://doi.org/10.1002/leap.1210.
43. Li, V. G., Dunn, M., Pearce, P., McCoy, D., Voelker, G. M., Savage, S., Levchenko, K., "Reading the tea leaves: A comparative analysis of threat intelligence," *Proceedings of the 28th USENIX Security Symposium*, pp. 851–867, 2019.
44. Chowdhary, K. R., "Natural Language Processing," in *Fundamentals of Artificial Intelligence*, K. Chowdhary, Ed., New Delhi: Springer India, 2020, pp. 603–649. https://doi.org/10.1007/978-81-322-3972-7_19.
45. Reimers, N., Gurevych, I., "Sentence-BERT: Sentence Embeddings using Siamese BERT-Networks," in *Proceedings of the 2019 Conference on Empirical Methods in Natural Language Processing and the 9th International Joint Conference on Natural Language Processing (EMNLP-IJCNLP)*, K. Inui, J. Jiang, V. Ng, and X. Wan, Eds., Hong Kong, China: Association for Computational Linguistics, Nov. 2019, pp. 3982–3992.https://doi.org/10.18653/v1/D19-1410.
46. Vaswani, A., Shazeer, N., Parmar, N., Uszkoreit, J., Jones, L., Gomez, A. N., Kaiser, L., Polosukhin, I., "Attention Is All You Need," Dec. 5, 2017. [Online]. Available: http://arxiv.org/abs/1706.03762.
47. Devlin, J., Chang, M.-W., Lee, K., Toutanova, K., "BERT: Pre-training of Deep Bidirectional Transformers for Language Understanding," *NORD'19*, vol. 1, pp. 4171–4186, Oct. 2018. https://doi.org/10.18653/v1/N19-1423.
48. Lee, J., Yoon, W., Kim, S., Kim, D., Kim, S., So, C. H., Kang, J., "BioBERT: A pre-trained biomedical language representation model for biomedical text mining," *Bioinformatics*, vol. 36, pp. 1234–1240, 2020. https://doi.org/10.1093/bioinformatics/btz682.
49. Mahdaouy, A. E., Lamsiyah, S., Idrissi, M. J., Alami, H., Yartaoui, Z., Berrada, I., "DomURLs_BERT: Pre-trained BERT-based Model for Malicious Domains and URLs Detection and Classification." (Sep. 13, 2024).
50. Tawil, Y., Alqaraleh, S., "BERT Based Topic-Specific Crawler," in *2021 Innovations in Intelligent Systems and Applications Conference (ASYU)*, Oct. 2021, pp. 1–5. https://doi.org/10.1109/ASYU52992.2021.9599076.

51. Apruzzese, G., Laskov, P., Tastemirova, A., "SoK: The Impact of Unlabelled Data in Cyberthreat Detection," in *2022 IEEE 7th European Symposium on Security and Privacy (EuroS&P)*, Jun. 2022, pp. 20–42. https://doi.org/10.1109/EuroSP53844.2022.00010.
52. Brown, T. B., Mann, B., Ryder, N., Subbiah, M., Kaplan, J., Dhariwal, P., Neelakantan, A., Shyam, P., Sastry, G., Askell, A., Agarwal, S., Herbert- Voss, A., Krueger, G., Henighan, T., Child, R., Ramesh, A., Ziegler, D. M., Wu, J., Winter, C., Hesse, C., Chen, M., Sigler, E., Litwin, M., Gray, S., Chess, B., Clark, J., Berner, C., McCandlish, S., Radford, A., Sutskever, I., Amodei, D., "Language Models are Few-Shot Learners," Jul. 22, 2020. [Online]. Available: http://arxiv.org/abs/2005.14165.
53. Lewis, P., Perez, E., Piktus, A., Petroni, F., Karpukhin, V., Goyal, N., Küttler, H., Lewis, M., Yih, W.-t., Rocktäschel, T., Riedel, S., Kiela, D., "Retrieval-Augmented Generation for Knowledge-Intensive NLP Tasks," in *Advances in Neural Information Processing Systems*, vol. 33, Curran Associates, Inc., 2020, pp. 9459–9474. [Online]. Available: https://proceedings.neurips.cc/paper/2020/hash/6b493230205f780e1bc26945df7481e5-Abstract.html.
54. Rahali, A., Akhloufi, M. A., "MalBERT: Using Transformers for Cybersecurity and Malicious Software Detection," Mar. 5, 2021. [Online]. Available: http://arxiv.org/abs/2103.03806.
55. Du, P.-Y., Zhang, N., Ebrahimi, M., Samtani, S., Lazarine, B., Arnold, N., Dunn, R., Suntwal, S., Angeles, G., Schweitzer, R., Chen, H., "Identifying, Collecting, and Presenting Hacker Community Data: Forums, IRC, Carding Shops, and DNMs," in *2018 IEEE International Conference on Intelligence and Security Informatics (ISI)*, Miami, FL: IEEE, Nov. 2018, pp. 70–75. https://doi.org/10.1109/ISI.2018.8587327.
56. Nunes, E., Diab, A., Gunn, A., Marin, E., Mishra, V., Paliath, V., Robertson, J., Shakarian, J., Thart, A., Shakarian, P., "Darknet and deepnet mining for proactive cybersecurity threat intelligence," in *2016 IEEE International Conference on Intelligence and Security Informatics (ISI)*, Tucson, AZ, USA: IEEE, Sep. 2016, pp. 7–12. https://doi.org/10.1109/ISI.2016.7745435.
57. Husari, G., Al-Shaer, E., Ahmed, M., Chu, B., Niu, X., "TTPDrill: Automatic and Accurate Extraction of Threat Actions from Unstructured Text of CTI Sources," in *Proceedings of the 33rd Annual Computer Security Applications Conference*, New York, NY, USA: Association for Computing Machinery, Dec. 4, 2017, pp. 103–115. https://doi.org/10.1145/3134600.3134646.
58. Du, M., Li, F., Zheng, G., Srikumar, V., "DeepLog: Anomaly Detection and Diagnosis from System Logs through Deep Learning," in *Proceedings of the 2017 ACM SIGSAC Conference on Computer and Communications Security*, Dallas Texas USA: ACM, Oct. 30, 2017, pp. 1285–1298. https://doi.org/10.1145/3133956.3134015.
59. Alhogail, A., Alsabih, A., "Applying machine learning and natural language processing to detect phishing email," *Computers & Security*, vol. 110, p. 102 414, Nov. 1, 2021. https://doi.org/10.1016/j.cose.2021.102414.
60. Sarker, I. H., Furhad, M. H., Nowrozy, R., "AI-Driven Cybersecurity: An Overview, Security Intelligence Modeling and Research Directions," *SN Computer Science*, vol. 2, p. 173, Mar. 26, 2021. https://doi.org/10.1007/s42979-021-00557-0.
61. Fernandes, G., Rodrigues, J. J. P. C., Carvalho, L. F., Al-Muhtadi, J. F., ProenCca, M. L., "A comprehensive survey on network anomaly detection," *Telecommunication Systems*, vol. 70, pp. 447–489, Mar. 1, 2019. https://doi.org/10.1007/s11235-018-0475-8.

62. Cui, L., Qu, Y., Xie, G., Zeng, D., Li, R., Shen, S., Yu, S., "Security and Privacy-Enhanced Federated Learning for Anomaly Detection in IoT Infrastructures," *IEEE Transactions on Industrial Informatics*, vol. 18, pp. 3492–3500, 2022. https://doi.org/10.1109/TII.2021.3107783.
63. 63] Sharmeen, S., Huda, S., Abawajy, J., Hassan, M. M., "An adaptive framework against android privilege escalation threats using deep learning and semi-supervised approaches," *Applied Soft Computing*, vol. 89, p. 106 089, Apr. 1, 2020. https://doi.org/10.1016/j.asoc.2020.106089.
64. Monshizadeh, M., Naldurg, P., Venkatakrishnan, V. N., "MACE: Detecting Privilege Escalation Vulnerabilities in Web Applications," in *Proceedings of the 2014 ACM SIGSAC Conference on Computer and Communications Security*, New York, NY, USA: Association for Computing Machinery, Nov. 3, 2014, pp. 690–701. https://doi.org/10.1145/2660267.2660337.
65. Ferrag, M. A., Ndhlovu, M., Tihanyi, N., Cordeiro, L. C., Debbah, M., Lestable, T., Thandi, N. S., "Revolutionizing Cyber Threat Detection With Large Language Models: A Privacy-Preserving BERTBased Lightweight Model for IoT/IIoT Devices," *IEEE Access*, vol. 12, pp. 23 733–23 750, 2024. https://doi.org/10.1109/ACCESS.2024.3363469.
66. Awasthi, A., Bdair, M., Kumar, A. N., Thapa, S., S, N. D., Kumar, B. R., "NLP for Sentiment Analysis in Social Media Posts to Detect Suspicious Behaviour," in *2024 International Conference on Intelligent Algorithms for Computational Intelligence Systems (IACIS)*, Aug. 2024, pp. 1–6. https://doi.org/10.1109/IACIS61494.2024.10721839.
67. Ampel, B., Samtani, S., Zhu, H., Ullman, S., Chen, H., "Labeling Hacker Exploits for Proactive Cyber Threat Intelligence: A Deep Transfer Learning Approach," in *2020 IEEE International Conference on Intelligence and Security Informatics (ISI)*, Arlington, VA, USA: IEEE, Nov. 2020, pp. 1–6. https://doi.org/10.1109/ISI49825.2020.9280548.
68. Samtani, S., Zhu, H., Chen, H., "Proactively Identifying Emerging Hacker Threats from the Dark Web: A Diachronic Graph Embedding Framework (D-GEF)," *ACM Transactions on Privacy and Security*, vol. 23, pp. 1–33, 2020. https://doi.org/10.1145/3409289.
69. Jin, Y., Jang, E., Cui, J., Chung, J.-W., Lee, Y., Shin, S., "DarkBERT: A Language Model for the Dark Side of the Internet." (May 18, 2023).
70. Pise, A. A., Singh, S., K, H., Gadilkar, S., Esther, Z. B., Pise, G. S., Imuede, J., "Adapting to Evolving Cyber Threat Landscapes with Dynamic Security Protocol Management in Large-Scale IoT Sensor Networks," *International Journal of Wireless and Ad Hoc Communication*, pp. 25–40, Issue 2 Jan. 1, 2023. https://doi.org/10.54216/IJWAC.070202.
71. Agorbia-Atta, C., Atalor, I., Agyei, R. K., Nachinaba, R., "Leveraging AI and ML for Next-Generation Cloud Security: Innovations in Risk-Based Access Management," *World Journal of Advanced Research and Reviews*, vol. 23, pp. 1487–1497, 2024. https://doi.org/10.30574/wjarr.2024.23.3.2788.
72. Hammad, A. A., Ahmed, S. R., Abdul-Hussein, M. K., Ahmed, M. R., Majeed, D. A., Algburi, S., "Deep Reinforcement Learning for Adaptive Cyber Defense in Network Security," in *Proceedings of the Cognitive Models and Artificial Intelligence Conference*, New York, NY, USA: Association for Computing Machinery, Jun. 23, 2024, pp. 292–297. https://doi.org/10.1145/3660853.3660930.

73. Landauer, M., Skopik, F., Wurzenberger, M., Rauber, A., "Dealing with Security Alert Flooding: Using Machine Learning for Domain-independent Alert Aggregation," *ACM Transactions on Privacy and Security*, vol. 25, 18:1-18:36, Apr. 9, 2022. https://doi.org/10.1145/3510581.
74. Belabbess, B., Bairat, M., Lhez, J., Curé, O., "Combining Machine Learning and Semantics for Anomaly Detection," in *Knowledge Engineering and Knowledge Management*, C. Faron Zucker, C. Ghidini, A. Napoli, and Y. Toussaint, Eds., Cham: Springer International Publishing, 2018, pp. 503–518. https://doi.org/10.1007/978-3-030-03667-6_32.
75. Shmueli, G., Bruce, P. C., Yahav, I., Patel, N. R., Lichtendahl, K. C., *Data Mining for Business Analytics: Concepts, Techniques, and Applications in R*, 1st ed. Wiley Publishing, Aug. 2017.
76. Jothi, N., Rashid, N. A., Husain, W., "Data Mining in Healthcare—A Review," *Procedia Computer Science*, vol. 72, pp. 306–313, Jan. 1, 2015.https://doi.org/10.1016/j.procs.2015.12.145.
77. Buczak, A. L., Guven, E., "A Survey of Data Mining and Machine Learning Methods for Cyber Security Intrusion Detection," *IEEE Communications Surveys & Tutorials*, vol. 18, pp. 1153–1176, 2016. https://doi.org/10.1109/COMST.2015.2494502.
78. Fouché, G., Langit, L., "Introduction to Data Mining," in *Foundations of SQL Server 2008 R2 Business Intelligence*, G. Fouché and L. Langit, Eds., Berkeley, CA: Apress, 2011, pp. 369–402. https://doi.org/10.1007/978-1-4302-3325-1_14.
79. Manning, C., Raghavan, P., Schuetze, H., *Introduction to Information Retrieval*. Cambridge University Press, 2008. [Online]. Available: https://www-nlp.stanford.edu/IR-book/.
80. Croft, W. B., Metzler, D., Strohman, T., *Search Engines: Information Retrieval in Practice*, 1st ed. USA: Addison-Wesley Publishing Company, 2009, 542 pp.
81. Aggarwal, C. C., "Knowledge-Based Recommender Systems," in *Recommender Systems: The Textbook*, C. C. Aggarwal, Ed., Cham: Springer International Publishing, 2016, pp. 167–197.https://doi.org/10.1007/978-3-319-29659-3_5.
82. Bouraga, S., Jureta, I., Faulkner, S., Herssens, C., "Knowledge-Based Recommendation Systems: A Survey," *International Journal of Intelligent Information Technologies (IJIIT)*, vol. 10, pp. 1–19, Apr. 1, 2014.https://doi.org/10.4018/ijiit.2014040101.
83. Nie, J.-Y., *Cross-Language Information Retrieval*. Cham: Springer International Publishing, 2010. https://doi.org/10.1007/978-3-031-02138-1.
84. Piplai, A., Mittal, S., Joshi, A., Finin, T., Holt, J., Zak, R., "Creating Cybersecurity Knowledge Graphs From Malware After Action Reports," *IEEE Access*, vol. 8, pp. 211 691–211 703, 2020. https://doi.org/10.1109/ACCESS.2020.3039234.
85. Han, X., Yuan, S., Trabelsi, M., "LogGPT: Log Anomaly Detection via GPT," in *2023 IEEE International Conference on Big Data (BigData)*, Dec. 2023, pp. 1117–1122. https://doi.org/10.1109/BigData59044.2023.10386543.
86. Le, T. H. M., Chen, H., Babar, M. A., "A Survey on Data-driven Software Vulnerability Assessment and Prioritization," *ACM Computing Surveys*, vol. 55, 100:1-100:39, Dec. 3, 2022. https://doi.org/10.1145/3529757.
87. Pan, J., Liang, W. S., Yidi, Y., "RAGLog: Log Anomaly Detection using Retrieval Augmented Generation," in *2024 IEEE World Forum on Public Safety Technology (WFPST)*, May 2024, pp. 169–174. https://doi.org/10.1109/WFPST58552.2024.00034.
88. Saba, T., Rehman, A., Sadad, T., Kolivand, H., Bahaj, S. A., "Anomalybased intrusion detection system for IoT networks through deep learning model," *Computers and*

Electrical Engineering, vol. 99, p. 107 810, Apr. 1, 2022. https://doi.org/10.1016/j.compeleceng.2022.107810.
89. Hossain Faruk, M. J., Shahriar, H., Valero, M., Barsha, F. L., Sobhan, S., Khan, M. A., Whitman, M., Cuzzocrea, A., Lo, D., Rahman, A., Wu, F., "Malware Detection and Prevention using Artificial Intelligence Techniques," in *2021 IEEE International Conference on Big Data (Big Data)*, Dec. 2021, pp. 5369–5377. https://doi.org/10.1109/BigData52589.2021.9671434.
90. Bansal, B., Jenipher, V. N., Jain, R., Dilip, R., Kumbhkar, M., Pramanik, S., Roy, S., Gupta, A., "Big Data Architecture for Network Security," in *Cyber Security and Network Security*, John Wiley & Sons, Ltd, 2022, pp. 233–267. https://doi.org/10.1002/9781119812555.ch11.
91. Abbasi, M., Shahraki, A., Taherkordi, A., "Deep Learning for Network Traffic Monitoring and Analysis (NTMA): A Survey," *Computer Communications*, vol. 170, pp. 19–41, Mar. 15, 2021. https://doi.org/10.1016/j.comcom.2021.01.021.
92. González-Granadillo, G., González-Zarzosa, S., Diaz, R., "Security Information and Event Management (SIEM): Analysis, Trends, and Usage in Critical Infrastructures," *Sensors*, vol. 21, p. 4759, 2021. https://doi.org/10.3390/s21144759.
93. Basheer, R., Alkhatib, B., "Threats from the Dark: A Review over Dark Web Investigation Research for Cyber Threat Intelligence," *Journal of Computer Networks and Communications*, vol. 2021, Z. Xu, Ed., pp. 1–21, Dec. 2021. https://doi.org/10.1155/2021/1302999.
94. Nazah, S., Huda, S., Abawajy, J., Hassan, M. M., "Evolution of Dark Web Threat Analysis and Detection: A Systematic Approach," *IEEE Access*, vol. 8, pp. 171 796–171 819, 2020. https://doi.org/10.1109/ACCESS.2020.3024198.
95. Bouwman, X., Griffioen, H., Egbers, J., Doerr, C., Klievink, B., Van Eeten, M., "A different cup of TI? the added value of commercial threat intelligence," in *Proceedings of the 29th USENIX Conference on Security Symposium*, USA: USENIX Association, Aug. 12, 2020, pp. 433–450.
96. Bouwman, X., Pochat, V. L., Foremski, P., Goethem, T. V., Gañán, C. H., Moura, G. C. M., Tajalizadehkhoob, S., Joosen, W., Eeten, M., "Helping hands: Measuring the impact of a large threat intelligence sharing community," 2022, pp. 1149–1165. [Online]. Available: https://www.usenix.org/conference/usenixsecurity22/presentation/bouwman.
97. Naumann, F., Rolker, C., "Assessment Methods for Information Quality Criteria," in *Fifth Conference on Information Quality (IQ 2000)*, MIT, 2000, pp. 148–162. https://doi.org/10.18452/2441.
98. NIST, "NVD Program Announcement," National Vulnerability Database. (Apr. 25, 2024), [Online]. Available: https://nvd.nist.gov/general/news/nvd-program-transition-announcement.
99. Kapko, M. "What's going on with the National Vulnerability Database?" Cybersecurity Dive. (Apr. 10, 2024), [Online]. Available: https://www.cybersecuritydive.com/news/nist-national-vulnerabilitydatabase/712826/.
100. Zorz, Z. "NIST's NVD has encountered a problem," Help Net Security. (Mar. 19, 2024), [Online]. Available: https://www.helpnetsecurity.com/2024/03/19/nvd-vulnerability-management/.
101. Kreutzer, D. M., Scheel, K., *Das governance-framework 5v*, Fraunhofer- Institut für Sichere Informationstechnologie, 2022. [Online]. Available: https://www.sit.fraunhofer.de/fileadmin/dokumente/studien_und_technical_reports/Governance-Framework_5V.pdf?_=1679318278.

102. Reinhold, T., *Towards a Peaceful Development of Cyberspace: Challenges and Technical Measures for the De-Escalation of State-Led Cyberconflicts and Arms Control of Cyberweapons*. Wiesbaden, Germany: Springer Vieweg, 2024. https://doi.org/10.1007/978-3-658-43951-4.
103. Haunschild, J., *Enhancing Citizens' Role in Public Safety: Interaction, Perception and Design of Mobile Warning Apps*. Wiesbaden, Germany: Springer Vieweg, 2025.
104. Führmann, D., Martín, L., Sánchez, L., Damer, N., "Anomaly Detection in Smart Environments: A Comprehensive Survey," *IEEE Access*, vol. 12, pp. 64 006–64 049, 2024. https://doi.org/10.1109/ACCESS.2024.3395051.
105. Bayer, M., *Deep Learning in Textual Low-Data Regimes for Cybersecurity*. Wiesbaden, Germany: Springer Vieweg, 2025.
106. McCloskey, M., Cohen, N. J., "Catastrophic Interference in Connectionist Networks: The Sequential Learning Problem," in *Psychology of Learning and Motivation*, G. H. Bower, Ed., vol. 24, Academic Press, Jan. 1, 1989, pp. 109–165. https://doi.org/10.1016/S0079-7421(08)60536-8.
107. Karantzas, G., Patsakis, C., "An empirical assessment of endpoint detection and response systems against advanced persistent threats attack vectors," *Journal of Cybersecurity and Privacy*, vol. 1, pp. 387–421, 2021.
108. Kilpatrick, L., Mallick, S. B., Kofman, R., "Gemini 1.5 Pro 2M context window, code execution capabilities, and Gemma 2 are available today," Gemini 1.5 Pro 2M context window, code execution capabilities, and Gemma 2 are available today. (Jun. 27, 2024), [Online]. Available: https://developers.googleblog.com/en/new-features-for-the-geminiapi-and-google-ai-studio/.
109. DeepSeek-AI, Guo, D., Yang, D., "DeepSeek-R1: Incentivizing Reasoning Capability in LLMs via Reinforcement Learning." (Jan. 22, 2025).
110. Calma, J. "Twitter just closed the book on academic research," The Verge. (May 31, 2023), [Online]. Available: https://www.theverge.com/2023/5/31/23739084/twitter-elon-musk-api-policy-chillingacademic-research.
111. Kaufhold, M.-A., Bayer, M., Reuter, C., "Rapid relevance classification of social media posts in disasters and emergencies: A system and evaluation featuring active, incremental and online learning," *Information Processing and Management*, vol. 57, pp. 102 132–102 132, 2020. https://doi.org/10.1016/j.ipm.2019.102132.
112. Reuter, C., Kaufhold, M.-A., "Fifteen Years of Social Media in Emergencies: A Retrospective Review and Future Directions for Crisis Informatics," *Journal of Contingencies and Crisis Management (JCCM)*, vol. 26, pp. 41–57, 2018. DOI: https://doi.org/10.1111/1468-5973.12196.
113. Simon, H. A., "Designing organizations for an information-rich world," *Computers, Communication, and the Public Interest*, vol. 70, pp. 187–202, 1971.
114. Srinivas Sajja, P., *Intelligent technologies for web applications*. Boca Raton, FL: CRC Press, 2012, 1 p.
115. Page, L., Brin, S., Motwani, R., Winograd, T., "The PageRank Citation Ranking: Bringing Order to the Web," Aug. 18, 1997.
116. Thiesen, J., Christensen, T. S., Kristensen, T. G., Andersen, R. D., Brunoe, B., Gregersen, T. K., Thrane, M., Weidema, B. P., "Rebound effects of price differences," *The International Journal of Life Cycle Assessment*, vol. 13, pp. 104–114, Mar. 1, 2008. https://doi.org/10.1065/lca2006.12.297.

117. Greenberg, A., "Inside a Firewall Vendor's 5-Year War With the Chinese Hackers Hijacking Its Devices," *Wired*, Oct. 31, 2024. [Online]. Available: https://www.wired.com/story/sophos-chengdu-china-five-yearhacker-war/.
118. Hanley, Z. "Palo Alto Expedition: From N-Day to Full Compromise," Horizon3.ai. (Oct. 9, 2024), [Online]. Available: https://www.horizon3.ai/attack-research/disclosures/palo-alto-expedition-from-nday-to-full-compromise/.
119. Schirrmacher, D. "Another critical vulnerability in Solarwind's Web Help Desk closed," heise Security. (Aug. 22, 2024), [Online]. Available: https://www.heise.de/en/news/Another-critical-vulnerabilityin-Solarwind-s-Web-Help-Desk-closed-9843872.html.
120. Fortinet, "Missing authentication in fgfmsd," FortiGuard Labs. (Oct. 23, 2024), [Online]. Available: https://fortiguard.com/psirt/FG-IR-24-423.
121. Knop, D. "Cisco reports more than 35 vulnerabilities in firewall products," heise Security. (Oct. 24, 2024), [Online]. Available: https://www.heise.de/en/news/Cisco-reports-more-than-35-vulnerabilities-infirewall-products-9992758.html.
122. Knop, D. "Ivanti patcht Endpoint Manager, Avalanche, VPN- und NACSoftware," heise Security. (Nov. 13, 2024), [Online]. Available: https://www.heise.de/en/news/Ivanti-patcht-Endpoint-Manager-Avalanche-VPN-und-NAC-Software-10031698.html.
123. Thomson, I. "GCHQ cyber-chief slams security outfits peddling 'medieval witchcraft'," The Register. (Feb. 3, 2017), [Online]. Available: https://www.theregister.com/2017/02/03/security_threat_solutions/.
124. McQuade, M., "The Untold Story of NotPetya, the Most Devastating Cyberattack in History," *Wired*, pp. 1–6, 2018. [Online]. Available: https://www.wired.com/story/notpetya-cyberattack-ukraine-russiacode-crashed-the-world/.
125. Sabottke, C., Suciu, O., Dumitras, T., "Vulnerability Disclosure in the Age of Social Media: Exploiting Twitter for Predicting Real-World Exploits," in *24th USENIX Security Symposium (USENIX Security 15)*, Washington, D.C.: USENIX Association, Aug. 2015, pp. 1041–1056. [Online]. Available: https://www.usenix.org/conference/usenixsecurity15/technical-sessions/presentation/sabottke.
126. Riebe, T., Kaufhold, M.-A., Reuter, C., "The Impact of Organizational Structure and Technology Use on Collaborative Practices in Computer Emergency Response Teams: An Empirical Study," *Proceedings of the ACM on Human-Computer Interaction*, vol. 5, pp. 1–30, CSCW2 Oct. 13, 2021. https://doi.org/10.1145/3479865.
127. Stieglitz, S., Mirbabaie, M., Fromm, J., Melzer, S., "The adoption of social media analytics for crisis management—Challenges and opportunities," in *26th European Conference on Information Systems: Beyond Digitization—Facets of Socio-Technical Change, ECIS 2018*, 2018.
128. Zade, H., Shah, K., Rangarajan, V., Kshirsagar, P., Imran, M., Starbird, K., "From Situational Awareness to Actionability: Towards Improving the Utility of Social Media Data for Crisis Response," in *Proceedings of the ACM on Human-Computer Interaction*, 2018.
129. Reuter, C., Ludwig, T., Kaufhold, M.-A., Pipek, V., "XHELP: Design of a cross-platform social-media application to support volunteermoderators in disasters," in *Conference on Human Factors in Computing Systems—Proceedings*, B. Begole, J. Kim, K. Inkpen, and W. Woo, Eds., vol. 2015- April, Seoul: ACM Press, 2015, pp. 4093–4102. https://doi.org/10.1145/2702123.2702171.
130. Ruefle, R., Dorofee, A., Mundie, D., Householder, A. D., Murray, M., Perl, S. J., "Computer Security Incident Response Team Development and Evolution," *IEEE Security & Privacy*, vol. 12, pp. 16–26, 2014. https://doi.org/10.1109/MSP.2014.89.

131. Endsley, M. R., "Towards a theory of situation awareness in dynamic systems," *Human Factors*, vol. 37, pp. 32–64, 1995. https://doi.org/10.1518/001872095779049543.
132. Husák, M., Jirsík, T., Yang, S. J., "SoK: Contemporary Issues and Challenges to Enable Cyber Situational Awareness for Network Security," in *Proceedings of the 15th International Conference on Availability, Reliability and Security*, New York, NY, USA: Association for Computing Machinery, 2020. https://doi.org/10.1145/3407023.3407062.
133. Franke, U., Brynielsson, J., "Cyber situational awareness-a systematic review of the literature," *Computers & Security*, vol. 46, pp. 18–31, 2014.
134. Gutzwiller, R., Dykstra, J., Payne, B., "Gaps and Opportunities in Situational Awareness for Cybersecurity," *Digital Threats*, vol. 1, Sep. 2020. https://doi.org/10.1145/3384471.
135. Drodt, M., Pagel, L., Biedorf, T., "Einbindung Datenschutz und Betriebsrat beim Aufbau eines SIEM," in *Cybersecurity Best Practices*, Springer, 2018, pp. 271–284.
136. Matta, L., Husák, M., "A Dashboard for Cyber Situational Awareness and Decision Support in Network Security Management," in *2021 IFIP/IEEE International Symposium on Integrated Network Management (IM)*, 2021, pp. 716–717.
137. F. Skopik, T. Páhi, and M. Leitner, Eds., *Cyber Situational Awareness in Public-Private-Partnerships: Organisationsübergreifende Cyber- Sicherheitsvorfälle effektiv bewältigen*. Berlin, Heidelberg: Springer Berlin Heidelberg, 2018. https://doi.org/10.1007/978-3-662-56084-6.
138. Booth, H., Rike, D., Witte, G., "The National Vulnerability Database: Overview," 2013, pp. 3–3. [Online]. Available: https://ws680.nist.gov/publication/get_pdf.cfm?pub_id=915172.
139. Lekkas, D., Spinellis, D., "Handling and reporting security advisories: A scorecard approach," *IEEE Security Privacy*, vol. 3, pp. 32–41, 2005. https://doi.org/10.1109/MSP.2005.98.
140. Mavroeidis, V., Bromander, S., "Cyber Threat Intelligence Model: An Evaluation of Taxonomies, Sharing Standards, and Ontologies within Cyber Threat Intelligence," in *2017 European Intelligence and Security Informatics Conference (EISIC)*, Sep. 2017, pp. 91–98. https://doi.org/10.1109/EISIC.2017.20.
141. Plohmann, D., Clauß, M., Enders, S., Padilla, E., "Malpedia: A Collaborative Effort to Inventorize the Malware Landscape," *The Journal on Cybercrime & Digital Investigations*, vol. 3, pp. 1–19, Dec. 2017. [Online]. Available: https://journal.cecyf.fr/ojs/index.php/cybin/article/view/17.
142. Wagner, C., Dulaunoy, A., Wagener, G., Iklody, A., "MISP: The Design and Implementation of a Collaborative Threat Intelligence Sharing Platform," in *Proceedings of the 2016 ACM on Workshop on Information Sharing and Collaborative Security*, New York, NY, USA: Association for Computing Machinery, 2016, pp. 49–56. https://doi.org/10.1145/2994539.2994542.
143. Abuse.ch, *ThreatFox: Share Indicators of Compromise*, 2022. [Online]. Available: https://threatfox.abuse.ch/.
144. Pulsedive, *Pulsedive: Threat Intelligence Made Easy*, 2022. [Online]. Available: https://pulsedive.com/.
145. Mittal, S., Das, P. K., Mulwad, V., Joshi, A., Finin, T., "CyberTwitter: Using Twitter to generate alerts for cybersecurity threats and vulnerabilities," in *2016 IEEE/ACM International Conference on Advances in Social Networks Analysis and Mining (ASONAM)*, IEEE, Aug. 2016, pp. 860–867. https://doi.org/10.1109/ASONAM.2016.7752338.

146. Kaufhold, M.-A., *Information Refinement Technologies for Crisis Informatics: User Expectations and Design Principles for Social Media and Mobile Apps*. Wiesbaden, Germany: Springer Vieweg, 2021, 381 p. 52 illus. https://doi.org/10.1007/978-3-658-33341-6.
147. Rodriguez, A., Okamura, K., "Generating Real Time Cyber Situational Awareness Information Through Social Media Data Mining," in *2019 IEEE 43rd Annual Computer Software and Applications Conference (COMPSAC)*, vol. 2, Jul. 2019, pp. 502–507. https://doi.org/10.1109/COMPSAC.2019.10256.
148. Montesino, R., Fenz, S., Baluja, W., "SIEM-based framework for security controls automation," *Information Management & Computer Security*, vol. 20, pp. 248–263, Jan. 1, 2012. https://doi.org/10.1108/09685221211267639.
149. Mounika, V., Yuan, X., Bandaru, K., "Analyzing CVE Database Using Unsupervised Topic Modelling," in *2019 International Conference on Computational Science and Computational Intelligence (CSCI)*, Dec. 2019, pp. 72–77. https://doi.org/10.1109/CSCI49370.2019.00019.
150. Nyanchama, M., "Enterprise Vulnerability Management and Its Role in Information Security Management.," *Information Systems Security*, vol. 14, pp. 29–56, 2005.
151. Chalvatzis, I., Karras, D. A., Papademetriou, R. C., "Evaluation of Security Vulnerability Scanners for Small and Medium Enterprises Business Networks Resilience towards Risk Assessment," in *2019 IEEE International Conference on Artificial Intelligence and Computer Applications (ICAICA)*, Mar. 2019, pp. 52–58. https://doi.org/10.1109/ICAICA.2019.8873438.
152. Scarfone, K. A., Mell, P. M., "Guide to Intrusion Detection and Prevention Systems (IDPS)," National Institute of Standards and Technology, Gaithersburg, MD, 2007, NIST SP 800-94. https://doi.org/10.6028/NIST.SP.800-94.
153. Scarfone, K., Mell, P., "Intrusion Detection and Prevention Systems," in *Handbook of Information and Communication Security*, P. Stavroulakis and M. Stamp, Eds., Berlin, Heidelberg: Springer, 2010, pp. 177–192. https://doi.org/10.1007/978-3-642-04117-4_9.
154. Klett, G., Kersten, H., Schröder, K.-W., *IT-Notfallmanagement mit System*. Springer, 2011.
155. Williams, C., Donaldson, S., Siegel, S., *Building an Effective Security Program*. De Gruyter, Sep. 21, 2020.https://doi.org/10.1515/9781501506529.
156. Hassan, W. U., Bates, A., Marino, D., "Tactical Provenance Analysis for Endpoint Detection and Response Systems," in *2020 IEEE Symposium on Security and Privacy (SP)*, May 2020, pp. 1172–1189. https://doi.org/10.1109/SP40000.2020.00096.
157. Tahboub, R., Saleh, Y., "Data Leakage/Loss Prevention Systems (DLP)," in 2014 World Congress on Computer Applications and Information Systems (WCCAIS), 2014, pp. 1–6. https://doi.org/10.1109/WCCAIS.2014.6916624.
158. Tundis, A., Ruppert, S., Mühlhüuser, M., "On the Automated Assessment of Open-Source Cyber Threat Intelligence Sources," in *Computational Science—ICCS 2020*, V. V. Krzhizhanovskaya, G. Závodszky, M. H. Lees, J. J. Dongarra, P. M. A. Sloot, S. Brissos, and J. Teixeira, Eds., Cham: Springer International Publishing, 2020, pp. 453–467. https://doi.org/10.1007/978-3-030-50417-5_34.
159. Hassan, N. A., Hijazi, R., "Social Media Intelligence," in *Open Source Intelligence Methods and Tools*, Springer, 2018, pp. 203–260.https://doi.org/10.1007/978-1-4842-3213-2_5.

160. Lykou, G., Anagnostopoulou, A., Stergiopoulos, G., Gritzalis, D., "Cybersecurity Self-assessment Tools: Evaluating the Importance for Securing Industrial Control Systems in Critical Infrastructures," in *Critical Information Infrastructures Security*, E. Luiijf, I. Žutautait?e, and B. M. Hümmerli, Eds., Cham: Springer International Publishing, 2019, pp. 129–142.https://doi.org/10.1007/978-3-030-05849-4_10.
161. Couch, A. L., "System configuration management," in *Handbook of Network and System Administration*, Elsevier, 2008, pp. 75–133.https://doi.org/10.1016/B978-044452198-9.50006-9.
162. Flexera, *Software Vulnerability Management. Stop reacting. Gain control. Stay secure.* 2022. [Online]. Available: https://www.flexera.com/products/software-vulnerability-manager.
163. Corporation, I. B. M., *IBM Security QRadar. Visibility, detection, investigation, and response*, 2021. [Online]. Available: https://www.ibm.com/downloads/cas/OP62GKAR.
164. Corporation, T. M., *Situation Awareness*, 2021. [Online]. Available: https://www.mitre.org/capabilities/cybersecurity/situation-awareness#.
165. Shankaranarayanan, G., Iyer, B., Stoddard, D. B., "Quality of Social Media Data and Implications of Social Media for Data Quality," *undefined*, 2012. [Online]. Available: https://www.semanticscholar.org/paper/Quality-of-Social-Media-Data-and-Implications-of-Shankaranarayanan-Iyer/4db13709cb106fb65be16bf85316c5806fc40241.
166. Rohweder, J. P., Kasten, G., Malzahn, D., Piro, A., Schmid, J., "Informationsqualitüt—definitionen, dimensionen und begriffe," in *Datenund Informationsqualitüt: Auf dem Weg zur Information Excellence*, K. Hildebrand, M. Gebauer, H. Hinrichs, and M. Mielke, Eds., Wiesbaden: Vieweg+Teubner, 2011, pp. 25–45. https://doi.org/10.1007/978-3-8348-9953-8_2.
167. Moh, M., Pininti, S., Doddapaneni, S., Moh, T.-S., "Detecting web attacks using multi-stage log analysis," in *2016 IEEE 6th international conference on advanced computing (IACC)*, IEEE, 2016, pp. 733–738.
168. Kritikos, K., Magoutis, K., Papoutsakis, M., Ioannidis, S., "A survey on vulnerability assessment tools and databases for cloud-based web applications," *Array*, vol. 3-4, p. 100 011, Sep. 1, 2019.https://doi.org/10.1016/j.array.2019.100011.
169. You, W., Zong, P., Chen, K., Wang, X., Liao, X., Bian, P., Liang, B., "SemFuzz: Semantics-based Automatic Generation of Proof-of-Concept Exploits," in *Proceedings of the 2017 ACM SIGSAC Conference on Computer and Communications Security*, New York, NY, USA: Association for Computing Machinery, Oct. 30, 2017, pp. 2139–2154. https://doi.org/10.1145/3133956.3134085.
170. Jyothsna, V. V. R. P. V., Prasad, R., Prasad, K. M., "A review of anomaly based intrusion detection systems," *International Journal of Computer Applications*, vol. 28, pp. 26–35, 2011.
171. Ou, Y.-j., Lin, Y., Zhang, Y., "The design and implementation of hostbased intrusion detection system," in *2010 third international symposium on intelligent information technology and security informatics*, IEEE, 2010, pp. 595–598.
172. Kumar, V., Sangwan, O. P., "Signature based intrusion detection system using SNORT," *International Journal of Computer Applications & Information Technology*, vol. 1, pp. 35–41, 2012.

173. Kaur, K., Gupta, I., Singh, A. K., "A comparative evaluation of data leakage/ loss prevention systems (DLPS)," in *Proc. 4th Int. Conf. Computer Science & Information Technology (CS & IT-CSCP)*, 2017, pp. 87–95.
174. Alneyadi, S., Sithirasenan, E., Muthukkumarasamy, V., "A survey on data leakage prevention systems," *Journal of Network and Computer Applications*, vol. 62, pp. 137–152, 2016.
175. Chen, H., Liu, R., Park, N., Subrahmanian, V. S., "Using twitter to predict when vulnerabilities will be exploited," in *Proceedings of the ACM SIGKDD International Conference on Knowledge Discovery and Data Mining*, New York, New York, USA: ACM Press, 2019, pp. 3143–3152. https://doi.org/10.1145/3292500.3330742.
176. Chen, H., Liu, J., Liu, R., Park, N., Subrahmanian, V., "VASE: A Twitter- Based Vulnerability Analysis and Score Engine," in *2019 IEEE International Conference on Data Mining (ICDM)*, Nov. 2019, pp. 976–981. https://doi.org/10.1109/ICDM.2019.00110.
177. Lyu, X., Ding, Y., Yang, S.-H., "Safety and security risk assessment in cyber-physical systems," *IET Cyber-Physical Systems: Theory & Applications*, vol. 4, pp. 221–232, 2019.
178. Venkatasubramanian, K. K., Nabar, S., Gupta, S., Poovendran, R., "Cyber physical security solutions for pervasive health monitoring systems," in *User-Driven Healthcare*, vol. 1, IGI Global, Jan. 31, 2013, pp. 447–465. https://doi.org/10.4018/978-1-4666-2770-3.ch022.
179. Schmittner, C., Ma, Z., Puschner, P., "Limitation and Improvement of STPA-Sec for Safety and Security Co-analysis," in *Computer Safety, Reliability, and Security*, A. Skavhaug, J. Guiochet, E. Schoitsch, and F. Bitsch, Eds., Cham: Springer International Publishing, 2016, pp. 195–209. https://doi.org/10.1007/978-3-319-45480-1_16.
180. Aksu, M. U., Dilek, M. H., Tatli, E. I., Bicakci, K., Dirik, H. I., Demirezen, M. U., Aykir, T., "A quantitative CVSS-based cyber security risk assessment methodology for IT systems," in *2017 International Carnahan Conference on Security Technology (ICCST)*, IEEE, 2017, pp. 1–8.
181. Kure, H. I., Islam, S., Razzaque, M. A., "An integrated cyber security risk management approach for a cyber-physical system," *Applied Sciences*, vol. 8, p. 898, 2018.
182. Mayring, P., "Qualitative Content Analysis," *Forum: Qualitative Social Research*, vol. 1, 2000.
183. Strauss, A. L., Corbin, J., *Basics of qualitative research: Techniques and procedures for developing grounded theory*. Sage Publications, 1998. [Online]. Available: https://us.sagepub.com/en-us/nam/basics-ofqualitative-research/book235578.
184. Kaufhold, M.-A., Stöttinger, M., Reuter, C., "Cyber Threat Observatory: Desing and Evaluation of an Interactive Dashboard for Computer Emergency Response Teams," in *Proceedings of the European Conference on Information Systems (ECIS)*, 2022.
185. Habdank, M., Rodehutskors, N., Koch, R., "Relevancy assessment of tweets using supervised learning techniques: Mining emergency related tweets for automated relevancy classification," in *2017 4th International conference on information and communication technologies for disaster management (ICT-DM)*, IEEE, 2017, pp. 1–8.
186. Widup, S., Hylender, D., Bassett, G., Langlois, P., Pinto, A., *2020 Verizon Data Breach Investigations Report*. May 19, 2020.https://doi.org/10.13140/RG.2.2.21300.48008.
187. Eun, S. T., "Cyber warfare in the russo-ukrainian war: Assessment and implications," *Institute of Foreign Affairs and National Security*, vol. 16, pp. 1–4, 2022. [Online]. Available: https://kiss.kstudy.com/Detail/Ar?key=3959268.

188. Soman, B., "Black Hat USA 2019," Aug. 8, 2019. [Online]. Available: https://www.blackhat.com/us-19/briefings/schedule/index.html#death-to-the-ioc-whats-next-in-threat-intelligence-15392.
189. Wagner, T. D., Mahbub, K., Palomar, E., Abdallah, A. E., "Cyber threat intelligence sharing: Survey and research directions," *Computers & Security*, vol. 87, p. 101 589, Nov. 1, 2019. https://doi.org/10.1016/j.cose.2019.101589.
190. Tounsi, W., Rais, H., "A survey on technical threat intelligence in the age of sophisticated cyber attacks," *Computers & Security*, vol. 72, pp. 212–233, Jan. 1, 2018. https://doi.org/10.1016/j.cose.2017.09.001.
191. Mikolov, T., Chen, K., Corrado, G., Dean, J., "Efficient Estimation of Word Representations in Vector Space," in *Proceedings of the International Conference on Learning Representations*, Jan. 16, 2013.
192. Gururangan, S., Marasovic, A., Swayamdipta, S., Lo, K., Beltagy, I., Downey, D., Smith, N. A., "Don't Stop Pretraining: Adapt Language Models to Domains and Tasks," in *Proceedings of the 58th Annual Meeting of the Association for Computational Linguistics*, Stroudsburg, PA, USA: Association for Computational Linguistics, 2020, pp. 8342–8360. https://doi.org/10.18653/v1/2020.acl-main.740.
193. Beltagy, I., Lo, K., Cohan, A., "SciBERT: A Pretrained Language Model for Scientific Text," Mar. 2019. [Online]. Available: http://arxiv.org/abs/1903.10676.
194. Alsentzer, E., Murphy, J., Boag, W., Weng, W.-H., Jindi, D., Naumann, T., McDermott, M., "Publicly Available Clinical BERT Embeddings," in *Proceedings of the 2nd Clinical Natural Language Processing Workshop*, A. Rumshisky, K. Roberts, S. Bethard, and T. Naumann, Eds., Minneapolis, Minnesota, USA: Association for Computational Linguistics, Jun. 2019, pp. 72–78. https://doi.org/10.18653/v1/W19-1909.
195. Mumtaz, S., Rodriguez, C., Benatallah, B., Al-Banna, M., Zamanirad, S., "Learning Word Representation for the Cyber Security Vulnerability Domain," in *Proceedings of the International Joint Conference on Neural Networks*, 2020. https://doi.org/10.1109/IJCNN48605.2020.9207140.
196. Lee, Y., Saxe, J., Harang, R., "CATBERT: Context-Aware Tiny BERT for Detecting Social Engineering Emails," Oct. 7, 2020. [Online]. Available: http://arxiv.org/abs/2010.03484.
197. Krieger, J.-D., Spinde, T., Ruas, T., Kulshrestha, J., Gipp, B., "A Domainadaptive Pre-training Approach for Language Bias Detection in News," in *Proceedings of the 22nd ACM/IEEE Joint Conference on Digital Libraries*, Jun. 20, 2022, pp. 1–7. https://doi.org/10.1145/3529372.3530932.
198. Liu, Y., Ott, M., Goyal, N., Du, J., Joshi, M., Chen, D., Levy, O., Lewis, M., Zettlemoyer, L., Stoyanov, V., "RoBERTa: A Robustly Optimized BERT Pretraining Approach." (Jul. 26, 2019).
199. Mendsaikhan, O., Hasegawa, H., Yamaguchi, Y., Shimada, H., Bataa, E., "Identification of Cybersecurity Specific Content Using Different Language Models," *Journal of Information Processing*, vol. 28, pp. 623–632, 2020. https://doi.org/10.2197/ipsjjip.28.623.
200. Yin, J., Tang, M., Cao, J., Wang, H., "Apply transfer learning to cybersecurity: Predicting exploitability of vulnerabilities by description," *Knowledge-Based Systems*, vol. 210, p. 106 529, Dec. 2020. https://doi.org/10.1016/j.knosys.2020.106529.
201. Souza, F., Nogueira, R., Lotufo, R., "Portuguese Named Entity Recognition using BERT-CRF." (Feb. 27, 2020).

202. Zhang, W., Jiang, S., Zhao, S., Hou, K., Liu, Y., Zhang, L., "A BERTBiLSTM- CRF Model for Chinese Electronic Medical Records Named Entity Recognition," in *2019 12th International Conference on Intelligent Computation Technology and Automation (ICICTA)*, Oct. 2019, pp. 166–169. https://doi.org/10.1109/ICICTA49267.2019.00043.
203. Bayer, M., Kaufhold, M.-A., Reuter, C., "Survey on Data Augmentation for Text Classification," *ACM Computing Surveys*, 2022. https://doi.org/10.1145/3544558.
204. Joulin, A., Grave, E., Bojanowski, P., Douze, M., Jégou, H., Mikolov, T., "FastText.zip: Compressing text classification models." (Dec. 12, 2016).
205. Kemker, R., McClure, M., Abitino, A., Hayes, T., Kanan, C., "Measuring Catastrophic Forgetting in Neural Networks," *Proceedings of the AAAI Conference on Artificial Intelligence*, vol. 32, Apr. 29, 2018. https://doi.org/10.1609/aaai.v32i1.11651.
206. Su, L., Guo, J., Zhang, R., Fan, Y., Lan, Y., Cheng, X., "Continual Domain Adaptation for Machine Reading Comprehension," in *Proceedings of the 29th ACM International Conference on Information & Knowledge Management*, New York, NY, USA: Association for Computing Machinery, Oct. 19, 2020, pp. 1395–1404. https://doi.org/10.1145/3340531.3412047.
207. Rongali, S., Jagannatha, A., Rawat, B. P. S., Yu, H., "Continual Domain- Tuning for Pretrained Language Models." (Mar. 19, 2021).
208. Sun, C., Qiu, X., Xu, Y., Huang, X., "How to Fine-Tune BERT for Text Classification?" In *Chinese Computational Linguistics*, M. Sun, X. Huang, H. Ji, Z. Liu, and Y. Liu, Eds., Cham: Springer International Publishing, 2019, pp. 194–206. https://doi.org/10.1007/978-3-030-32381-3_16.
209. Ibor, A. E., Okunoye, O. B., Oladeji, F. A., Abdulsalam, K. A., "Novel Hybrid Model for Intrusion Prediction on Cyber Physical Systems' Communication Networks based on Bio-inspired Deep Neural Network Structure," *Journal of Information Security and Applications*, vol. 65, p. 103 107, Mar. 1, 2022.https://doi.org/10.1016/j.jisa.2021.103107.
210. Kingma, D. P., Ba, J., "Adam: A Method for Stochastic Optimization," *CoRR*, Dec. 22, 2014. https://doi.org/10.48550/arXiv.1412.6980.
211. Bayer, M., Kaufhold, M.-A., Reuter, C., "Information Overload in Crisis Management: Bilingual Evaluation of Embedding Models for Clustering Social Media Posts in Emergencies," in *Proceedings of the European Conference on Information Systems (ECIS)*, Marrakech, Morocco, 2021, pp. 1–18. [Online]. Available: https://tuprints.ulb.tu-darmstadt.de/22167/.
212. Taylor, W. L., ""Cloze Procedure": A New Tool for Measuring Readability," *Journalism Quarterly*, vol. 30, pp. 415–433, Sep. 1, 1953. https://doi.org/10.1177/107769905303000401.
213. Wang, A., Pruksachatkun, Y., Nangia, N., Singh, A., Michael, J., Hill, F., Levy, O., Bowman, S. R., "SuperGLUE: A Stickier Benchmark for General-Purpose Language Understanding Systems," Feb. 12, 2020. [Online]. Available: http://arxiv.org/abs/1905.00537.
214. Reimers, N., Gurevych, I., "Reporting Score Distributions Makes a Difference: Performance Study of LSTM-networks for Sequence Tagging," in *Proceedings of the 2017 Conference on Empirical Methods in Natural Language Processing*, M. Palmer, R. Hwa, and S. Riedel, Eds., Copenhagen, Denmark: Association for Computational Linguistics, Sep. 2017, pp. 338–348. https://doi.org/10.18653/v1/D17-1035.

215. Sanh, V., Debut, L., Chaumond, J., Wolf, T., "DistilBERT, a distilled version of BERT: Smaller, faster, cheaper and lighter." (Feb. 29, 2020).
216. Tam, D., R. Menon, R., Bansal, M., Srivastava, S., Raffel, C., "Improving and Simplifying Pattern Exploiting Training," in *Proceedings of the 2021 Conference on Empirical Methods in Natural Language Processing*, M.-F. Moens, X. Huang, L. Specia, and S. W.-t. Yih, Eds., Online and Punta Cana, Dominican Republic: Association for Computational Linguistics, Nov. 2021, pp. 4980–4991. https://doi.org/10.18653/v1/2021.emnlp-main.407.
217. Xiang, G., Hong, J., Rose, C. P., Cranor, L., "CANTINA+: A Feature- Rich Machine Learning Framework for Detecting Phishing Web Sites," *ACM Transactions on Information and System Security*, vol. 14, 21:1- 21:28, Sep. 1, 2011. https://doi.org/10.1145/2019599.2019606.
218. Yang, P., Zhao, G., Zeng, P., "Phishing Website Detection Based on Multidimensional Features Driven by Deep Learning," *IEEE Access*, vol. 7, pp. 15 196–15 209, 2019. https://doi.org/10.1109/ACCESS.2019.2892066.
219. Salem, A., Banescu, S., Pretschner, A., "Maat: Automatically Analyzing VirusTotal for Accurate Labeling and Effective Malware Detection," *ACM Transactions on Privacy and Security*, vol. 24, 25:1-25:35, Jul. 19, 2021. https://doi.org/10.1145/3465361.
220. Shang, R., Zhu, S., Ren, J., Liu, H., Jiao, L., "Evolutionary neural architecture search based on evaluation correction and functional units," *Knowledge-Based Systems*, vol. 251, p. 109 206, Sep. 5, 2022. https://doi.org/10.1016/j.knosys.2022.109206.
221. Okunoye, O. B., Ibor, A. E., "Hybrid fake news detection technique with genetic search and deep learning," *Computers and Electrical Engineering*, vol. 103, p. 108 344, Oct. 1, 2022. https://doi.org/10.1016/j.compeleceng.2022.108344.
222. Ramnani, R. R., Shivaram, K., Sengupta, S., M. A. K., "Semi-Automated Information Extraction from Unstructured Threat Advisories," in *Proceedings of the 10th Innovations in Software Engineering Conference*, New York, NY, USA: Association for Computing Machinery, Feb. 5, 2017, pp. 181–187. https://doi.org/10.1145/3021460.3021482.
223. Le Sceller, Q., Karbab, E. B., Debbabi, M., Iqbal, F., "SONAR: Automatic Detection of Cyber Security Events over the Twitter Stream," in *ARES'17*, Reggio Calabria Italy: ACM, Aug. 29, 2017, pp. 1–11. https://doi.org/10.1145/3098954.3098992.
224. Elbaz, C., Rilling, L., Morin, C., "Fighting N-day vulnerabilities with automated CVSS vector prediction at disclosure," in *Proceedings of the 15th International Conference on Availability, Reliability and Security*, New York, NY, USA: ACM, Aug. 2020, pp. 1–10. https://doi.org/10.1145/3407023.3407038.
225. Chakrabarti, S., Berg, M., Dom, B., "Focused crawling: A new approach to topic-specific Web resource discovery," *Computer Networks*, vol. 31, pp. 1623–1640, 1999. https://doi.org/10.1016/S1389-1286(99)00052-3.
226. Peters, M. E., Neumann, M., Iyyer, M., Gardner, M., Clark, C., Lee, K., Zettlemoyer, L., "Deep Contextualized Word Representations," in *NAACL'18*, New Orleans, Louisiana: Association for Computational Linguistics, Jun. 2018, pp. 2227–2237. https://doi.org/10.18653/v1/N18-1202.
227. Wang, W., Chen, X., Zou, Y., Wang, H., Dai, Z., "A Focused Crawler Based on Naive Bayes Classifier," in *2010 Third International Symposium on Intelligent Information Technology and Security Informatics*, Apr. 2010, pp. 517–521. https://doi.org/10.1109/IITSI.2010.30.

228. Liu, W., He, Y., Wu, J., Du, Y., Liu, X., Xi, T., Gan, Z., Jiang, P., Huang, X., "A focused crawler based on semantic disambiguation vector space model," *Complex & Intelligent Systems*, vol. 9, pp. 345–366, Feb. 1, 2023. https://doi.org/10.1007/s40747-022-00707-8.
229. Pham, K., Santos, A., Freire, J., "Bootstrapping Domain-Specific Content Discovery on the Web," in *The World Wide Web Conference*, New York, NY, USA: Association for Computing Machinery, May 13, 2019, pp. 1476–1486. https://doi.org/10.1145/3308558.3313709.
230. Sanagavarapu, L. M., Mathur, N., Agrawal, S., Reddy, Y. R., "SIREN—Security Information Retrieval and Extraction eNgine," in *Advances in Information Retrieval*, G. Pasi, B. Piwowarski, L. Azzopardi, and A. Hanbury, Eds., Cham: Springer International Publishing, 2018, pp. 811–814. https://doi.org/10.1007/978-3-319-76941-7_81.
231. Tundis, A., Ruppert, S., Mühlhüuser, M., "A Feature-driven Method for Automating the Assessment of OSINT Cyber Threat Sources," *Computers & Security*, vol. 113, p. 102 576, Feb. 1, 2022.https://doi.org/10.1016/j.cose.2021.102576.
232. Dekel, L., Leybovich, I., Zilberman, P., Puzis, R., "MABAT: A Multi- Armed Bandit Approach for Threat-Hunting," *IEEE Transactions on Information Forensics and Security*, vol. 18, pp. 477–490, 2023. https://doi.org/10.1109/TIFS.2022.3215010.
233. Robert M Lee, "Cyber Threat Intelligence Survey." (Feb. 10, 2020), [Online]. Available: https://www.sans.org/white-papers/39395/.
234. Brown, R., Lee, R. M., "SANS Cyber Threat Intelligence Survey 2021," SANS, Jan. 18, 2021. [Online]. Available: https://www.sans.org/white-papers/40080/.
235. Brown, R., Stirparo, P., "SANS Cyber Threat Intelligence Survey 2022," SANS, Feb. 23, 2022. [Online]. Available: https://www.sans.org/whitepapers/sans-2022-cyber-threat-intelligence-survey/.
236. Auer, P., Cesa-Bianchi, N., Fischer, P., "Finite-time Analysis of the Multiarmed Bandit Problem," *Machine Learning*, vol. 47, pp. 235–256, May 1, 2002. https://doi.org/10.1023/A:1013689704352.
237. Pham, K., Santos, A., Freire, J., "Learning to Discover Domain-Specific Web Content," in *Proceedings of the Eleventh ACM International Conference on Web Search and Data Mining*, New York, NY, USA: Association for Computing Machinery, Feb. 2, 2018, pp. 432–440. https://doi.org/10.1145/3159652.3159724.
238. Chakrabarti, S., Gibson, D. A., McCurley, K. S., "Surfing the Web backwards," *Computer Networks*, vol. 31, pp. 1679–1693, May 17, 1999. https://doi.org/10.1016/S1389-1286(99)00042-0.
239. Diligenti, M., Coetzee, F., Lawrence, S., Giles, C. L., Gori, M., "Focused Crawling Using Context Graphs," in *Proceedings of the 26th International Conference on Very Large Data Bases*, San Francisco, CA, USA: Morgan Kaufmann Publishers Inc., Sep. 10, 2000, pp. 527–534.
240. Grootendorst, M., *KeyBERT*, version v0.1.3, Zenodo, Jan. 25, 2021. [Online]. Available: https://zenodo.org/record/4461265.
241. Amur, Z. H., Hooi, Y. K., Soomro, G. M., Bhanbhro, H., Karyem, S., Sohu, N., "Unlocking the Potential of Keyword Extraction: The Need for Access to High-Quality Datasets," *Applied Sciences*, vol. 13, p. 7228, 2023. https://doi.org/10.3390/app13127228.
242. Fleck, A. "Infographic: Cybercrime Expected To Skyrocket in Coming Years," Statista Daily Data. (Feb. 22, 2024), [Online]. Available: https://www.statista.com/chart/28878/expected-cost-of-cybercrimeuntil-2027.

243. Fachkha, C., Debbabi, M., "Darknet as a Source of Cyber Intelligence: Survey, Taxonomy, and Characterization," *IEEE Communications Surveys & Tutorials*, vol. 18, pp. 1197–1227, Sum. 2016. https://doi.org/10.1109/COMST.2015.2497690.
244. Dong, F., Yuan, S., Ou, H., Liu, L., "New Cyber Threat Discovery from Darknet Marketplaces," in *2018 IEEE Conference on Big Data and Analytics (ICBDA)*, Langkawi, Malaysia: IEEE, Nov. 2018, pp. 62–67. https://doi.org/10.1109/ICBDAA.2018.8629658.
245. Hernández, I., Rivero, C. R., Ruiz, D., "Deep Web crawling: A survey," *World Wide Web*, vol. 22, pp. 1577–1610, 2019. https://doi.org/10.1007/s11280-018-0602-1.
246. Schafer, M., Fuchs, M., Strohmeier, M., Engel, M., Liechti, M., Lenders, V., "BlackWidow: Monitoring the Dark Web for Cyber Security Information," in *2019 11th International Conference on Cyber Conflict (CyCon)*, Tallinn, Estonia: IEEE, May 2019, pp. 1–21. https://doi.org/10.23919/CYCON.2019. 8756845.
247. Marin, E., Almukaynizi, M., Sarkar, S., Nunes, E., Shakarian, J., Shakarian, P., *Exploring Malicious Hacker Communities: Toward Proactive Cyber-Defense*. Cambridge: Cambridge University Press, 2021. https://doi.org/10.1017/9781108869003.
248. D. Harris, Ed., *Engineering Psychology and Cognitive Ergonomics: 12th International Conference, EPCE 2015, Held as Part of HCI International 2015, Los Angeles, CA, USA, August 2-7, 2015, Proceedings*. Cham: Springer International Publishing, 2015, vol. 9174. https://doi.org/10.1007/978-3-319-20373-7.
249. Pastor-Galindo, J., Nespoli, P., Gómez Mármol, F., Martínez Pérez, G., "The Not Yet Exploited Goldmine of OSINT: Opportunities, Open Challenges and Future Trends," *IEEE Access*, vol. 8, pp. 10 282–10 304, 2020. https://doi.org/10.1109/ACCESS.2020.2965257.
250. Bergman, J., Popov, O. B., "Exploring Dark Web Crawlers: A Systematic Literature Review of Dark Web Crawlers and Their Implementation," *IEEE Access*, vol. 11, pp. 35 914–35 933, 2023. https://doi.org/10.1109/ACCESS.2023.3255165.
251. Benjamin, V., Valacich, J. S., Chen, H., "DICE-E: A Framework for Conducting Darknet Identification, Collection, Evaluation with Ethics," *MIS Quarterly*, vol. 43, pp. 1–22, Jan. 1, 2019. https://doi.org/10.25300/MISQ/2019/13808.
252. Hughes, J., Pastrana, S., Hutchings, A., Afroz, S., Samtani, S., Li, W., Santana Marin, E., "The Art of Cybercrime Community Research," *ACM Computing Surveys*, vol. 56, 155:1-155:26, Feb. 23, 2024. https://doi.org/10.1145/3639362.
253. Sobhan, S., Williams, T., Faruk, M. J. H., Rodriguez, J., Tasnim, M., Mathew, E., Wright, J., Shahriar, H., "A Review of Dark Web: Trends and Future Directions," in *2022 IEEE 46th Annual Computers, Software, and Applications Conference (COMPSAC)*, Los Alamitos, CA, USA: IEEE, Jun. 2022, pp. 1780–1785. https://doi.org/10.1109/COMPSAC54236.2022.00283.
254. Cherqi, O., Mezzour, G., Ghogho, M., El Koutbi, M., "Analysis of Hacking Related Trade in the Darkweb," in *2018 IEEE International Conference on Intelligence and Security Informatics (ISI)*, Nov. 2018, pp. 79–84. https://doi.org/10.1109/ISI.2018.8587311.
255. Deliu, I., Leichter, C., Franke, K., "Extracting cyber threat intelligence from hacker forums: Support vector machines versus convolutional neural networks," in *2017 IEEE International Conference on Big Data*, Boston, MA: IEEE, Dec. 2017, pp. 3648–3656. https://doi.org/10.1109/BigData.2017.8258359.

256. Kadoguchi, M., Hayashi, S., Hashimoto, M., Otsuka, A., "Exploring the Dark Web for Cyber Threat Intelligence using Machine Leaning," in *2019 IEEE International Conference on Intelligence and Security Informatics (ISI)*, Shenzhen, China: IEEE, Jul. 2019, pp. 200–202. https://doi.org/10.1109/ISI.2019.8823360.
257. Koloveas, P., Chantzios, T., Tryfonopoulos, C., Skiadopoulos, S., "A Crawler Architecture for Harvesting the Clear, Social, and Dark Web for IoT-Related Cyber-Threat Intelligence," in *2019 IEEE World Congress on Services (SERVICES)*, Milan, Italy: IEEE, Jul. 2019, pp. 3–8. https://doi.org/10.1109/SERVICES.2019.00016.
258. Deliu, I., Leichter, C., Franke, K., "Collecting Cyber Threat Intelligence from Hacker Forums via a Two-Stage, Hybrid Process using Support Vector Machines and Latent Dirichlet Allocation," in *2018 IEEE International Conference on Big Data*, Seattle, WA, USA: IEEE, Dec. 2018, pp. 5008–5013. https://doi.org/10.1109/BigData.2018.8622469.
259. Marin, E., Almukaynizi, M., Nunes, E., Shakarian, P., "Community Finding of Malware and Exploit Vendors on Darkweb Marketplaces," in *2018 1st International Conference on Data Intelligence and Security (ICDIS)*, South Padre Island, TX, USA: IEEE, Apr. 2018, pp. 81–84. https://doi.org/10.1109/ICDIS.2018.00019.
260. Sapienza, A., Bessi, A., Damodaran, S., Shakarian, P., Lerman, K., Ferrara, E., "Early-Warnings of Cyber Threats in Online Discussions," in *2017 IEEE International Conference on Data Mining Workshops (ICDMW)*, New Orleans, LA: IEEE, Nov. 2017, pp. 667–674. https://doi.org/10.1109/ICDMW.2017.94.
261. Almukaynizi, M., Grimm, A., Nunes, E., Shakarian, J., Shakarian, P., "Predicting Cyber Threats through Hacker Social Networks in Darkweb and Deepweb Forums," in *Proceedings of the 2017 International Conference of The Computational Social Science Society of the Americas*, Santa Fe NM USA: ACM, Oct. 19, 2017, pp. 1–7. https://doi.org/10.1145/3145574.3145590.
262. Sarkar, S., Almukaynizi, M., Shakarian, J., Shakarian, P., "Predicting enterprise cyber incidents using social network analysis on dark web hacker forums," *The Cyber Defense Review*, vol. CYCON U.S. pp. 87–102, 2018. [Online]. Available: https://www.jstor.org/stable/26846122.
263. Zulkarnine, A. T., Frank, R., Monk, B., Mitchell, J., Davies, G., "Surfacing collaborated networks in dark web to find illicit and criminal content," in *2016 IEEE International Conference on Intelligence and Security Informatics (ISI)*, Tucson, AZ, USA: IEEE, Sep. 2016, pp. 109–114. https://doi.org/10.1109/ISI.2016.7745452.
264. Hayes, D., Cappa, F., Cardon, J., "A Framework for More Effective Dark Web Marketplace Investigations," *Information-an International Interdisciplinary Journal*, vol. 9, p. 186, 2018. https://doi.org/10.3390/info9080186.
265. Williams, R., Samtani, S., Patton, M., Chen, H., "Incremental Hacker Forum Exploit Collection and Classification for Proactive Cyber Threat Intelligence: An Exploratory Study," in *2018 IEEE International Conference on Intelligence and Security Informatics (ISI)*, Miami, FL: IEEE, Nov. 2018, pp. 94–99. https://doi.org/10.1109/ISI.2018.8587336.
266. Samtani, S., Chinn, R., Chen, H., Nunamaker, J. F., "Exploring Emerging Hacker Assets and Key Hackers for Proactive Cyber Threat Intelligence," *Journal of Management Information Systems*, vol. 34, pp. 1023–1053, Oct. 2, 2017. https://doi.org/10.1080/07421222.2017.1394049.

267. Grisham, J., Samtani, S., Patton, M., Chen, H., "Identifying mobile malware and key threat actors in online hacker forums for proactive cyber threat intelligence," in *2017 IEEE International Conference on Intelligence and Security Informatics (ISI)*, Beijing, China: IEEE, Jul. 2017, pp. 13–18. https://doi.org/10.1109/ISI.2017.8004867.
268. Shin, G.-Y., Jang, Y., Kim, D.-W., Park, S., Park, A.-R., Kim, Y., Han, M.-M., "Dark Side of the Web: Dark Web Classification Based on TextCNN and Topic Modeling Weight," *IEEE Access*, vol. 12, pp. 36 361–36 371, 2024. https://doi.org/10.1109/ACCESS.2023.3347737.
269. Adewopo, V., Gonen, B., Elsayed, N., Ozer, M., Elsayed, Z. S., "Deep Learning Algorithm for Threat Detection in Hackers Forum (Deep Web)." (Feb. 3, 2022).
270. Pastrana, S., Thomas, D. R., Hutchings, A., Clayton, R., "CrimeBB: Enabling Cybercrime Research on Underground Forums at Scale," in *Proceedings of the 2018 World Wide Web Conference on World Wide Web—WWW '18*, Lyon, France: ACM Press, 2018, pp. 1845–1854. https://doi.org/10.1145/3178876.3186178.
271. Deguara, N., Arshad, J., Paracha, A., Azad, M. A., "Threat Miner—A Text Analysis Engine for Threat Identification Using Dark Web Data," in *2022 IEEE International Conference on Big Data (Big Data)*, Dec. 2022, pp. 3043–3052. https://doi.org/10.1109/BigData55660.2022.10020397.
272. Tavabi, N., Goyal, P., Almukaynizi, M., Shakarian, P., Lerman, K., "Dark- Embed: Exploit prediction with neural language models," in *Thirty-Second AAAI Conference on Artificial Intelligence*, New Orleans, Louisiana, USA: AAAI Press, Feb. 2, 2018, pp. 7849–7854. https://doi.org/10.1609/aaai.v32i1.11428.
273. Nikoletos, S., Raftopoulou, P., "Employing social network analysis to dark web communities," in *2022 IEEE International Conference on Cyber Security and Resilience (CSR)*, Rhodes, Greece: IEEE, Jul. 27, 2022, pp. 311–316. https://doi.org/10.1109/CSR54599.2022.9850325.
274. Ebrahimi, M., Surdeanu, M., Samtani, S., Chen, H., "Detecting Cyber Threats in Non-English Dark Net Markets: A Cross-Lingual Transfer Learning Approach," in 2018 IEEE International Conference on Intelligence and Security Informatics (ISI), Miami, FL: IEEE, Nov. 2018, pp. 85–90. https://doi.org/10.1109/ISI.2018.8587404.
275. Marin, E., Almukaynizi, M., Shakarian, P., "Reasoning About Future Cyber-Attacks Through Socio-Technical Hacking Information," in *2019 IEEE 31st International Conference on Tools with Artificial Intelligence (ICTAI)*, Portland, OR, USA: IEEE, Nov. 2019, pp. 157–164. https://doi.org/10.1109/ICTAI.2019.00030.
276. Almukaynizi, M., Nunes, E., Dharaiya, K., Senguttuvan, M., Shakarian, J., Shakarian, P., "Proactive identification of exploits in the wild through vulnerability mentions online," in *2017 International Conference on Cyber Conflict (CyCon U.S.)*, Washington, DC: IEEE, Nov. 2017, pp. 82–88. https://doi.org/10.1109/CYCONUS.2017.8167501.
277. Robertson, J., Diab, A., Marin, E., Nunes, E., Paliath, V., Shakarian, J., Shakarian, P., "Darknet Mining and Game Theory for Enhanced Cyber Threat Intelligence," *The Cyber Defense Review*, vol. 1, pp. 95–122, 2016. [Online]. Available: https://www.jstor.org/stable/26267362.
278. Samtani, S., Chai, Y., Chen, H., "Linking Exploits from the Dark Web to Known Vulnerabilities for Proactive Cyber Threat Intelligence: An Attention-Based Deep Structured Semantic Model," *MIS Quarterly*, vol. 46, pp. 911–946, Jun. 1, 2022. [Online]. Available: https://aisel.aisnet.org/misq/vol46/iss2/11.

279. Fu, T., Abbasi, A., Chen, H., "A focused crawler for Dark Web forums," *Journal of the American Society for Information Science and Technology*, vol. 61, pp. 1213–1231, 2010. https://doi.org/10.1002/asi.21323.
280. Al Nabki, M. W., Fidalgo, E., Alegre, E., De Paz, I., "Classifying Illegal Activities on Tor Network Based on Web Textual Contents," in *Proceedings of the 15th Conference of the European Chapter of the Association for Computational Linguistics: Volume 1, Long Papers*, Valencia, Spain: Association for Computational Linguistics, 2017, pp. 35–43. https://doi.org/10.18653/v1/E17-1004.
281. Yannikos, Y., Heeger, J., Steinebach, M., "Data Acquisition on a Large Darknet Marketplace," in *ARES '22: Proceedings of the 17th International Conference on Availability, Reliability and Security*, Vienna Austria: ACM, Aug. 2022, pp. 1–6. https://doi.org/10.1145/3538969.3544472.
282. Iliou, C., Kalpakis, G., Tsikrika, T., Vrochidis, S., Kompatsiaris, I., "Hybrid focused crawling on the Surface and the Dark Web," *EURASIP Journal on Information Security*, vol. 2017, p. 11, 2017. https://doi.org/10.1186/s13635-017-0064-5.
283. Park, J., Mun, H., Lee, Y., "Improving Tor Hidden Service Crawler Performance," in *2018 IEEE Conference on Dependable and Secure Computing (DSC)*, Kaohsiung, Taiwan: IEEE, Dec. 2018, pp. 1–8. https://doi.org/10.1109/DESEC.2018.8625103.
284. Samtani, S., Li, W., Benjamin, V., Chen, H., "Informing Cyber Threat Intelligence through Dark Web Situational Awareness: The AZSecure Hacker Assets Portal," *Digital Threats*, vol. 2, pp. 1–10, 2021. https://doi.org/10.1145/3450972.
285. Alkhatib, B., Basheer, R. S., "Mining the Dark Web: A Novel Approach for Placing a Dark Website under Investigation," *International Journal of Modern Education and Computer Science*, vol. 11, pp. 1–13, 2019. https://doi.org/10.5815/ijmecs.2019.10.01.
286. Baravalle, A., Lopez, M. S., Lee, S. W., "Mining the Dark Web: Drugs and Fake Ids," in *2016 IEEE 16th International Conference on Data Mining Workshops (ICDMW)*, Barcelona, Spain: IEEE, Dec. 2016, pp. 350–356. https://doi.org/10.1109/ICDMW.2016.0056.
287. Wu, Y., Zhao, F., Chen, X., Skums, P., Sevigny, E. L., Maimon, D., Ouellet, M., Swahn, M. H., Strasser, S. M., Feizollahi, M. J., Zhang, Y., Sekhon, G., "Python Scrapers for Scraping Cryptomarkets on Tor," in *Security, Privacy, and Anonymity in Computation, Communication, and Storage*, G. Wang, J. Feng, M. Z. A. Bhuiyan, and R. Lu, Eds., vol. 11611, Cham: Springer International Publishing, 2019, pp. 244–260. https://doi.org/10.1007/978-3-030-24907-6_19.
288. Xu, Y., Chen, G., Wu, J., Xu, W., Liu, Q., "Research on Dark Web Monitoring Crawler Based on TOR," in *2021 IEEE 2nd International Conference on Information Technology, Big Data and Artificial Intelligence (ICIBA)*, Chongqing, China: IEEE, Dec. 17, 2021, pp. 197–202.https://doi.org/10.1109/ICIBA52610.2021.9687954.
289. Ebrahimi, M., Nunamaker, J. F., Chen, H., "Semi-Supervised Cyber Threat Identification in Dark Net Markets: A Transductive and Deep Learning Approach," *Journal of Management Information Systems*, vol. 37, pp. 694–722, Jul. 2, 2020. https://doi.org/10.1080/07421222.2020.1790186.
290. Audran, D., Andersen, M., Hansen, M., Andersen, M., Frederiksen, T., Hansen, K., Georgoulias, D., Vasilomanolakis, E., "Tick Tock Break the Clock: Breaking CAPTCHAs on the Darkweb," in *Proceedings of the 19th International Conference on Security and Cryptography*, Lisbon, Portugal: SCITEPRESS—Science and Technology Publications, 2022, pp. 357–365. https://doi.org/10.5220/0011273300003283.

291. Celestini, A., Me, G., Mignone, M., "Tor Marketplaces Exploratory Data Analysis: The Drugs Case," in *Global Security, Safety and Sustainability—The Security Challenges of the Connected World*, H. Jahankhani, A. Carlile, D. Emm, A. Hosseinian-Far, G. Brown, G. Sexton, and A. Jamal, Eds., vol. 630, Cham: Springer International Publishing, 2016, pp. 218–229. https://doi.org/10.1007/978-3-319-51064-4_18.
292. Cuevas, A., Miedema, F., Soska, K., Christin, N., Wegberg, R., "Measurement by Proxy: On the Accuracy of Online Marketplace Measurements," in *31st USENIX Security Symposium (USENIX Security 22)*, 2022, pp. 2153–2170. [Online]. Available: https://www.usenix.org/conference/usenixsecurity22/presentation/cuevas.
293. Zhang, H., Zou, F., "A Survey of the Dark Web and Dark Market Research," in *2020 IEEE 6th International Conference on Computer and Communications (ICCC)*, Chengdu, China: IEEE, Dec. 11, 2020, pp. 1694–1705.https://doi.org/10.1109/ICCC51575.2020.9345271.
294. Fang, Y., Guo, Y., Huang, C., Liu, L., "Analyzing and Identifying Data Breaches in Underground Forums," *IEEE Access*, vol. 7, pp. 48 770–48 777, 2019. https://doi.org/10.1109/ACCESS.2019.2910229.
295. Zenebe, A., Shumba, M., Carillo, A., Cuenca, S., "Cyber Threat Discovery from Dark Web," in *Proceedings of 28th International Conference on Software Engineering and Data Engineering*, 2019, pp. 174–163. https://doi.org/10.29007/nkfk.
296. Arnold, N., Ebrahimi, M., Zhang, N., Lazarine, B., Patton, M., Chen, H., Samtani, S., "Dark-Net Ecosystem Cyber-Threat Intelligence (CTI) Tool," in *2019 IEEE International Conference on Intelligence and Security Informatics (ISI)*, Jul. 2019, pp. 92–97. https://doi.org/10.1109/ISI.2019.8823501.
297. Almukaynizi, M., Marin, E., Nunes, E., Shakarian, P., Simari, G. I., Kapoor, D., Siedlecki, T., "DARKMENTION: A Deployed System to Predict Enterprise-Targeted External Cyberattacks," in *2018 IEEE International Conference on Intelligence and Security Informatics (ISI)*, Miami, FL: IEEE, Nov. 2018, pp. 31–36. https://doi.org/10.1109/ISI.2018.8587334.
298. Georgoulias, D., Pedersen, J. M., Falch, M., Vasilomanolakis, E., "A qualitative mapping of Darkweb marketplaces," in *2021 APWG Symposium on Electronic Crime Research (eCrime)*, Dec. 2021, pp. 1–15. https://doi.org/10.1109/eCrime54498.2021.9738766.
299. Van Wegberg, R., Tajalizadehkhoob, S., Soska, K., Akyazi, U., Gañán, C., Klievink, B., Christin, N., Van Eeten, M., "Plug and prey? measuring the commoditization of cybercrime via online anonymous markets," in *27th USENIX Security Symposium (USENIX Security 18)*, USA: USENIX Association, Aug. 15, 2018, pp. 1009–1026.
300. Brenner, F., Platzer, F., Steinebach, M., "Discovery of Single-Vendor Marketplace Operators in the Tor-Network," in *ARES '21: Proceedings of the 16th International Conference on Availability, Reliability and Security*, New York, NY, USA: Association for Computing Machinery, Aug. 17, 2021, pp. 1–10. https://doi.org/10.1145/3465481.3470026.
301. Platzer, F., Schüfer, M., Steinebach, M., "Critical traffic analysis on the tor network," in *ARES '20: Proceedings of the 15th International Conference on Availability, Reliability and Security*, New York, NY, USA: Association for Computing Machinery, Aug. 25, 2020, pp. 1–10. https://doi.org/10.1145/3407023.3409180.
302. Steinebach, M., Schüfer, M., Karakuz, A., Brandl, K., Yannikos, Y., "Detection and Analysis of Tor Onion Services," in *ARES '19: Proceedings of the 14th International*

Conference on Availability, Reliability and Security, Canterbury CA United Kingdom: ACM, Aug. 2019, pp. 1–10. https://doi.org/10.1145/3339252.3341486.

303. Martin, J., Christin, N., “Ethics in cryptomarket research,” *International Journal of Drug Policy*, vol. 35, pp. 84–91, 2016. https://doi.org/10.1016/j.drugpo.2016.05.006.
304. Thomas, D. R., Pastrana, S., Hutchings, A., Clayton, R., Beresford, A. R., “Ethical issues in research using datasets of illicit origin,” in *Proceedings of the 2017 Internet Measurement Conference*, London United Kingdom: ACM, Nov. 2017, pp. 445–462. https://doi.org/10.1145/3131365.3131389.
305. Harviainen, J. T., Haasio, A., Ruokolainen, T., Hassan, L., Siuda, P., Hamari, J., “Information Protection in DarkWeb Drug Markets Research,” in *Proceedings of the 54th Hawaii International Conference on System Sciences*, 2021. https://doi.org/10.24251/HICSS.2021.567.
306. Platzer, F., Lux, A., “A Synopsis of Critical Aspects for Darknet Research,” in *Proceedings of the 17th International Conference on Availability, Reliability and Security*, Vienna Austria: ACM, Aug. 23, 2022, pp. 1–8. https://doi.org/10.1145/3538969.3544444.
307. Christin, N., “Traveling the silk road: A measurement analysis of a large anonymous online marketplace,” in *Proceedings of the 22nd international conference on World Wide Web*, Rio de Janeiro Brazil: ACM, May 13, 2013, pp. 213–224. https://doi.org/10.1145/2488388.2488408.
308. Cantelo, A., “Deeplight: Shining a Light on the Dark Web,” May 4, 2016, p. 12. [Online]. Available: https://www.slideshare.net/GavinOToole/deeplight-intelliagg.
309. Labrador, V., Pastrana, S., “Examining the trends and operations of modern Dark-Web marketplaces,” in *2022 IEEE European Symposium on Security and Privacy Workshops (EuroS&PW)*, Genoa, Italy: IEEE, Jun. 2022, pp. 163–172. https://doi.org/10.1109/EuroSPW55150.2022.00022.
310. Broséus, J., Rhumorbarbe, D., Morelato, M., Staehli, L., Rossy, Q., “A geographical analysis of trafficking on a popular darknet market,” *Forensic Science International*, vol. 277, pp. 88–102, 2017. https://doi.org/10.1016/j.forsciint.2017.05.021.
311. "The Dark Web as a Platform for Crime: An Exploration of Illicit Drug, Firearm, CSAM, and Cybercrime Markets,” in *The Palgrave Handbook of International Cybercrime and Cyberdeviance*, T. J. Holt and A. M. Bossler, Eds., Cham: Springer International Publishing, 2020, pp. 91–116. https://doi.org/10.1007/978-3-319-78440-3.
312. Samtani, S., Abate, M., Benjamin, V., Li, W., “Cybersecurity as an Industry: A Cyber Threat Intelligence Perspective,” in *The Palgrave Handbook of International Cybercrime and Cyberdeviance*, T. J. Holt and A. M. Bossler, Eds., Cham: Springer International Publishing, 2020, pp. 135–154. https://doi.org/10.1007/978-3-319-78440-3.
313. Ferry, N., Hackenheimer, T., Herrmann, F., Tourette, A., “Methodology of dark web monitoring,” in *2019 11th International Conference on Electronics, Computers and Artificial Intelligence (ECAI)*, Pitesti, Romania: IEEE, Jun. 2019, pp. 1–7. https://doi.org/10.1109/ECAI46879.2019.9042072.
314. Bernaschi, M., Celestini, A., Cianfriglia, M., Guarino, S., Lombardi, F., Mastrostefano, E., “Onion under Microscope: An in-depth analysis of the Tor Web,” *World Wide Web*, vol. 25, pp. 1287–1313, 2022. https://doi.org/10.1007/s11280-022-01044-z.
315. Lawrence, H., Hughes, A., Tonic, R., Zou, C., “D-miner: A framework for mining, searching, visualizing, and alerting on darknet events,” in *2017 IEEE Conference on*

Communications and Network Security (CNS), Las Vegas, NV: IEEE, Oct. 2017, pp. 1–9. https://doi.org/10.1109/CNS.2017.8228628.

316. Graczyk, M., Kinningham, K., "Automatic Product Categorization for Anonymous Marketplaces." (2015), [Online]. Available: https://cs229.stanford.edu/proj2015/184_report.pdf.
317. B. Akhgar, M. Gercke, S. Vrochidis, and H. Gibson, Eds., *Dark Web Investigation*. Cham: Springer International Publishing, 2021. https://doi.org/10.1007/978-3-030-55343-2.
318. Zhang, N., Ebrahimi, M., Li, W., Chen, H., "A Generative Adversarial Learning Framework for Breaking Text-Based CAPTCHA in the Dark Web," in *2020 IEEE International Conference on Intelligence and Security Informatics (ISI)*, Arlington, VA, USA: IEEE, Nov. 2020, pp. 1–6. https://doi.org/10.1109/ISI49825.2020.9280537.
319. Zhang, Y., Gao, H., Pei, G., Luo, S., Chang, G., Cheng, N., "A Survey of Research on CAPTCHA Designing and Breaking Techniques," in *2019 18th IEEE International Conference On Trust, Security And Privacy In Computing And Communications/13th IEEE International Conference On Big Data Science And Engineering (TrustCom/BigDataSE)*, Rotorua, New Zealand: IEEE, Aug. 2019, pp. 75–84. https://doi.org/10.1109/TrustCom/BigDataSE.2019.00020.
320. Csuka, K., Gaastra, D., *Breaking CAPTCHAs on the Dark Web*, System & Network Engineering—University of Amsterdam, 2018. [Online]. Available: https://www.os3.nl/_media/2017-2018/courses/rp1/p62_report.Pdf.
321. Weng, H., Zhao, B., Ji, S., Chen, J.,Wang, T., He, Q., Beyah, R., "Towards understanding the security of modern image captchas and underground captcha-solving services," *Big Data Mining and Analytics*, vol. 2, pp. 118–144, Jun. 2019. https://doi.org/10.26599/BDMA.2019.9020001.
322. Zhang, N., Ebrahimi, M., Li, W., Chen, H., "Counteracting Dark Web Text-Based CAPTCHA with Generative Adversarial Learning for Proactive Cyber Threat Intelligence," *ACM Transactions on Management Information Systems*, vol. 13, pp. 1–21, Jun. 30, 2022. https://doi.org/10.1145/3505226.
323. Nazah, S., Huda, S., Abawajy, J. H., Hassan, M. M., "An Unsupervised Model for Identifying and Characterizing Dark Web Forums," *IEEE Access*, vol. 9, pp. 112 871–112 892, 2021. https://doi.org/10.1109/ACCESS.2021.3103319.
324. Bullough, B. L., Yanchenko, A. K., Smith, C. L., Zipkin, J. R., "Predicting Exploitation of Disclosed Software Vulnerabilities Using Open-source Data," in *Proceedings of the 3rd ACM on International Workshop on Security And Privacy Analytics*, Scottsdale Arizona USA: ACM, Mar. 24, 2017, pp. 45–53. https://doi.org/10.1145/3041008.3041009.
325. Shor, P., "Algorithms for quantum computation: Discrete logarithms and factoring," in *Proceedings 35th Annual Symposium on Foundations of Computer Science*, Nov. 1994, pp. 124–134. https://doi.org/10.1109/SFCS.1994.365700.
326. C. Reuter, Ed., *Information Technology for Peace and Security*. Wiesbaden: Springer Fachmedien Wiesbaden, 2019. https://doi.org/10.1007/978-3-658-25652-4.
327. Perlroth, B. N., Scott, M., Frenkel, S., "Cyberattack Hits Ukraine Then Spreads Internationally," *The New York Times*, pp. 1–6, 2017. [Online]. Available: https://www.nytimes.com/2017/06/27/technology/ransomware-hackers.html.
328. Mu, D., Cuevas, A., Yang, L., Hu, H., Xing, X., Mao, B., Wang, G., "Understanding the reproducibility of crowd-reported security vulnerabilities," *Proceedings of the 27th USENIX Security Symposium*, pp. 919–936, 2018.

329. Chaparro, O., Lu, J., Zampetti, F., Moreno, L., Di Penta, M., Marcus, A., Bavota, G., Ng, V., "Detecting missing information in bug descriptions," *Proceedings of the ACM SIGSOFT Symposium on the Foundations of Software Engineering*, vol. Part F1301, pp. 396–407, 2017. https://doi.org/10.1145/3106237.3106285.
330. Fan, Y., Xia, X., Lo, D., Hassan, A. E., "Chaff from the Wheat: Characterizing and Determining Valid Bug Reports," *IEEE Transactions on Software Engineering*, vol. 46, pp. 495–525, 2018. https://doi.org/10.1109/TSE.2018.2864217.
331. Shahzad, M., Shafiq, M. Z., Liu, A. X., "A large scale exploratory analysis of software vulnerability life cycles," in *Proceedings—International Conference on Software Engineering*, IEEE, Jun. 2012, pp. 771–781. https://doi.org/10.1109/ICSE.2012.6227141.
332. Johnson, P., Lagerstrom, R., Ekstedt, M., Franke, U., "Can the common vulnerability scoring system be trusted? A Bayesian analysis," *IEEE Transactions on Dependable and Secure Computing*, vol. 15, pp. 1002–1015, 2018. https://doi.org/10.1109/TDSC.2016.2644614.
333. Akbik, A., Bergmann, T., Blythe, D., Rasul, K., Schweter, S., Vollgraf, R., "FLAIR: An Easy-to-Use Framework for State-of-the-Art NLP," in *Proceedings of the 2019 Conference of the North*, Stroudsburg, PA, USA: Association for Computational Linguistics, 2019, pp. 54–59. https://doi.org/10.18653/v1/N19-4010.
334. Nguyen, V. H., Massacci, F., "The (un)reliability of NVD vulnerable versions data: An empirical experiment on Google Chrome vulnerabilities," in *ASIA CCS 2013—Proceedings of the 8th ACM SIGSAC Symposium on Information, Computer and Communications Security*, New York, New York, USA: ACM Press, 2013, pp. 493–498. https://doi.org/10.1145/2484313.2484377.
335. Schaberreiter, T., Kupfersberger, V., Rantos, K., Spyros, A., Papanikolaou, A., Ilioudis, C., Quirchmayr, G., "A Quantitative Evaluation of Trust in the Quality of Cyber Threat Intelligence Sources," in *Proceedings of the 14th International Conference on Availability, Reliability and Security*, New York, NY, USA: ACM, Aug. 2019, pp. 1–10. https://doi.org/10.1145/3339252.3342112.
336. Farhang, S., Kirdan, M. B., Laszka, A., Grossklags, J., "An Empirical Study of Android Security Bulletins in Different Vendors," in *The Web Conference 2020—Proceedings of the World Wide Web Conference, WWW 2020*, New York, NY, USA: ACM, Apr. 2020, pp. 3063–3069.https://doi.org/10.1145/3366423.3380078.
337. Zhang, S., Ou, X., Caragea, D., "Predicting Cyber Risks through National Vulnerability Database," *Information Security Journal*, vol. 24, pp. 194–206, 2015. https://doi.org/10.1080/19393555.2015.1111961.
338. Guo, H., Xing, Z., Li, X., "Predicting Missing Information of Vulnerability Reports," in *Companion Proceedings of the Web Conference 2020*, New York, NY, USA: ACM, Apr. 2020, pp. 81–82. https://doi.org/10.1145/3366424.3382707.
339. Spring, J. M., Hatleback, E., Householder, A., Manion, A., Shick, D., "Towards Improving CVSS," *Cert.org*, Jun. 2018.
340. Fall, D., Kadobayashi, Y., "The Common Vulnerability Scoring System vs. Rock Star Vulnerabilities: Why the Discrepancy?" In *Proceedings of the 5th International Conference on Information Systems Security and Privacy*, SCITEPRESS—Science and Technology Publications, 2019, pp. 405–411. https://doi.org/10.5220/0007387704050411.
341. Holm, H., Afridi, K. K., "An expert-based investigation of the Common Vulnerability Scoring System," *Computers & Security*, vol. 53, pp. 18–30, 2015. https://doi.org/10.1016/j.cose.2015.04.012.

342. Allodi, L., Massacci, F., "Attack Potential in Impact and Complexity," in *Proceedings of the 12th International Conference on Availability, Reliability and Security*, Reggio Calabria, Italy: ACM, Aug. 2017, pp. 1–6. https://doi.org/10.1145/3098954.3098965.
343. Ruohonen, J., "A look at the time delays in CVSS vulnerability scoring," *Applied Computing and Informatics*, vol. 15, pp. 129–135, 2019. https://doi.org/10.1016/j.aci.2017.12.002.
344. Eppler, M. J., *Managing Information Quality*. Berlin, Heidelberg: Springer Berlin Heidelberg, 2003. https://doi.org/10.1007/978-3-540-24782-1.
345. Lample, G., Ballesteros, M., Subramanian, S., Kawakami, K., Dyer, C., "Neural architectures for named entity recognition," in *2016 Conference of the North American Chapter of the Association for Computational Linguistics: Human Language Technologies, NAACL HLT 2016—Proceedings of the Conference*, Stroudsburg, PA, USA: Association for Computational Linguistics, 2016, pp. 260–270. https://doi.org/10.18653/v1/n16-1030.
346. Huang, Z., Xu, W., Yu, K., "Bidirectional LSTM-CRF Models for Sequence Tagging," Aug. 9, 2015. [Online]. Available: http://arxiv.org/abs/1508.01991.
347. Chung, J., Gulcehre, C., Cho, K., Bengio, Y., "Empirical Evaluation of Gated Recurrent Neural Networks on Sequence Modeling," Dec. 11, 2014. [Online]. Available: http://arxiv.org/abs/1412.3555.
348. Smith, A., Osborne, M., "Using gazetteers in discriminative information extraction," *Proceedings of the Tenth Conference on Computational Natural Language Learning, CoNLL-X*, pp. 133–140, June 2006. https://doi.org/10.3115/1596276.1596302.
349. Krishnan, V., Ganapathy, V., *Named Entity Recognition*. 2005. [Online]. Available: http://www.stanford.edu/class/cs229/proj2005/KrishnanGanapathy-NamedEntityRecognition.pdf.
350. Wei, J., Zou, K., "EDA: Easy Data Augmentation Techniques for Boosting Performance on Text Classification Tasks," Aug. 25, 2019. [Online]. Available: http://arxiv.org/abs/1901.11196.
351. Elizabeth Dwoskin, Karla Adam, "Nations race to contain widespread hacking—TheWashington Post," *Washington Post*, May 13, 2017. [Online]. Available: https://www.washingtonpost.com/business/economy/nations-race-to-contain-widespread-hacking/2017/05/13/f6e5e992-3803-11e7-b412-62beef8121f7_story.html.
352. Brian Krebs, "A Basic Timeline of the Exchange Mass-Hack—Krebs on Security," Krebs on Security. (Mar. 8, 2021), [Online]. Available: https://krebsonsecurity.com/2021/03/a-basic-timeline-ofthe-exchange-mass-hack/.
353. Han, Z., Li, X., Xing, Z., Liu, H., Feng, Z., "Learning to Predict Severity of Software Vulnerability Using Only Vulnerability Description," in *2017 IEEE International Conference on Software Maintenance and Evolution (ICSME)*, Sep. 2017, pp. 125–136. https://doi.org/10.1109/ICSME.2017.52.
354. Shahid, M., Debar, H., "CVSS-BERT: Explainable Natural Language Processing to Determine the Severity of a Computer Security Vulnerability from its Description," Nov. 16, 2021. [Online]. Available: http://arxiv.org/abs/2111.08510.
355. Chen, H., Liu, J., Liu, R., Park, N., Subrahmanian, V. S., "VEST: A System for Vulnerability Exploit Scoring & Timing," pp. 6503–6505, 2019. [Online]. Available: https://www.ijcai.org/proceedings/2019/937.
356. Khazaei, A., Ghasemzadeh, M., Derhami, V., "An automatic method for CVSS score prediction using vulnerabilities description," *Journal of Intelligent and Fuzzy Systems*, vol. 30, pp. 89–96, 2016. https://doi.org/10.3233/IFS-151733.

357. Yamamoto, Y., Miyamoto, D., Nakayama, M., "Text-Mining Approach for Estimating Vulnerability Score," in *2015 4th International Workshop on Building Analysis Datasets and Gathering Experience Returns for Security (BADGERS)*, Nov. 2015, pp. 67–73. https://doi.org/10.1109/BADGERS.2015.018.
358. Blei, D. M., McAuliffe, J. D., "Supervised topic models," in *Proceedings of the 20th International Conference on Neural Information Processing Systems*, Red Hook, NY, USA: Curran Associates Inc., Dec. 3, 2007, pp. 121–128.
359. Breiman, L., "Random Forests," *Machine Learning*, vol. 45, pp. 5–32, Oct. 1, 2001. https://doi.org/10.1023/A:1010933404324.
360. Spanos, G., Angelis, L., Toloudis, D., "Assessment of vulnerability severity using text mining," in *ACM International Conference Proceeding Series*, vol. Part F1325, New York, New York, USA: ACM Press, 2017, pp. 1–6. https://doi.org/10.1145/3139367.3139390.
361. Freund, Y., Schapire, R. E., "A Short Introduction to Boosting," *Journal of Japanese Society for Artificial Intelligence*, vol. 14, pp. 771–780, 1999.
362. Gawron, M., Cheng, F., Meinel, C., "Automatic Vulnerability Classification Using Machine Learning," in *Risks and Security of Internet and Systems*, N. Cuppens, F. Cuppens, J.-L. Lanet, A. Legay, and J. Garcia- Alfaro, Eds., Cham: Springer International Publishing, 2018, pp. 3–17. https://doi.org/10.1007/978-3-319-76687-4_1.
363. Hochreiter, S., Schmidhuber, J., "Long Short-Term Memory," *Neural Computation*, vol. 9, pp. 1735–1780, Nov. 1, 1997.https://doi.org/10.1162/neco.1997.9.8.1735.
364. Sahin, S. E., Tosun, A., "A Conceptual Replication on Predicting the Severity of Software Vulnerabilities," in *Proceedings of the Evaluation and Assessment on Software Engineering*, New York, NY, USA: Association for Computing Machinery, Apr. 15, 2019, pp. 244–250. https://doi.org/10.1145/3319008.3319033.
365. Gong, X., Xing, Z., Li, X., Feng, Z., Han, Z., "Joint Prediction of Multiple Vulnerability Characteristics Through Multi-Task Learning," in *2019 24th International Conference on Engineering of Complex Computer Systems (ICECCS)*, Nov. 2019, pp. 31–40. https://doi.org/10.1109/ICECCS.2019.00011.
366. Liu, K., Zhou, Y., Wang, Q., Zhu, X., "Vulnerability Severity Prediction With Deep Neural Network," in *2019 5th International Conference on Big Data and Information Analytics (BigDIA)*, Jul. 2019, pp. 114–119. https://doi.org/10.1109/BigDIA.2019.8802851.
367. Jiang, Y., Atif, Y., "An Approach to Discover and Assess Vulnerability Severity Automatically in Cyber-Physical Systems," in *13th International Conference on Security of Information and Networks*, New York, NY, USA: Association for Computing Machinery, Feb. 1, 2021, pp. 1–8. https://doi.org/10.1145/3433174.3433612.
368. Yitagesu, S., Zhang, X., Feng, Z., Li, X., Xing, Z., "Automatic Partof- Speech Tagging for Security Vulnerability Descriptions," in *2021 IEEE/ACM 18th International Conference on Mining Software Repositories (MSR)*, May 2021, pp. 29–40. https://doi.org/10.1109/MSR52588.2021.00016.
369. Barbaresi, A., "Trafilatura: A Web Scraping Library and Command-Line Tool for Text Discovery and Extraction," in *ACL/IJCNLP'21*, Online: ACL, 2021, pp. 122–131. https://doi.org/10.18653/v1/2021.acl-demo.15.
370. Turc, I., Chang, M.-W., Lee, K., Toutanova, K., "Well-Read Students Learn Better: On the Importance of Pre-training Compact Models." (Sep. 25, 2019).

371. Wolf, T., Debut, L., Sanh, V., Chaumond, J., Delangue, C., Moi, A., Cistac, P., Rault, T., Louf, R., Funtowicz, M., Davison, J., Shleifer, S., Platen, P., Ma, C., Jernite, Y., Plu, J., Xu, C., Le Scao, T., Gugger, S., Drame, M., Lhoest, Q., Rush, A., "Transformers: State-of-the-Art Natural Language Processing," in *Proceedings of the 2020 Conference on Empirical Methods in Natural Language Processing: System Demonstrations*, Online: Association for Computational Linguistics, Oct. 2020, pp. 38–45. https://doi.org/10.18653/v1/2020.emnlp-demos.6.
372. Sundararajan, M., Taly, A., Yan, Q., "Axiomatic Attribution for Deep Networks," in *Proceedings of the 34th International Conference on Machine Learning*, PMLR, Jul. 17, 2017, pp. 3319–3328. [Online]. Available: https://proceedings.mlr.press/v70/sundararajan17a.html.
373. Preuveneers, D., Joosen, W., "Privacy-Preserving Polyglot Sharing and Analysis of Confidential Cyber Threat Intelligence," in *Proceedings of the 17th International Conference on Availability, Reliability and Security*, New York, NY, USA: Association for Computing Machinery, Aug. 23, 2022, pp. 1–11. https://doi.org/10.1145/3538969.3538982.
374. Skopik, F., Settanni, G., Fiedler, R., "A problem shared is a problem halved: A survey on the dimensions of collective cyber defense through security information sharing," *C&S*, vol. 60, pp. 154–176, 2016. https://doi.org/10.1016/j.cose.2016.04.003.
375. Gorzelak, K., Grudziecki, T., Jacewicz, P., Jaroszewski, P., Juszczyk, L., Kijews, P., "Proactive Detection of Network Security Incidents," European Union Agency for Cybersecurity, 2011.
376. Qader, W. A., Ameen, M. M., Ahmed, B. I., "An Overview of Bag of Words;Importance, Implementation, Applications, and Challenges," in *2019 International Engineering Conference (IEC)*, Jun. 2019, pp. 200–204. https://doi.org/10.1109/IEC47844.2019.8950616.
377. Harris, Z. S., "Distributional Structure," *WORD*, vol. 10, pp. 146–162, Aug. 1, 1954. https://doi.org/10.1080/00437956.1954.11659520.
378. Zong, C., Xia, R., Zhang, J., *Text Data Mining*. Singapore: Springer, 2021. https://doi.org/10.1007/978-981-16-0100-2.
379. Muennighoff, N., Tazi, N., Magne, L., Reimers, N., "MTEB: Massive Text Embedding Benchmark." (Mar. 19, 2023).
380. Ignaczak, L., Goldschmidt, G., Costa, C. A. D., Righi, R. D. R., "Text Mining in Cybersecurity: A Systematic Literature Review," *ACM Computing Surveys*, vol. 54, 140:1-140:36, Jul. 18, 2021. https://doi.org/10.1145/3462477.
381. Rousseeuw, P. J., "Silhouettes: A graphical aid to the interpretation and validation of cluster analysis," *Journal of Computational and Applied Mathematics*, vol. 20, pp. 53–65, Nov. 1, 1987. https://doi.org/10.1016/0377-0427(87)90125-7.
382. Lloyd, S., "Least squares quantization in PCM," *IEEE Transactions on Information Theory*, vol. 28, pp. 129–137, 1982. https://doi.org/10.1109/TIT.1982.1056489.
383. Ester, M., Kriegel, H.-P., Sander, J., Xu, X., "A density-based algorithm for discovering clusters in large spatial databases with noise," in *Proceedings of the Second International Conference on Knowledge Discovery and Data Mining*, Portland, Oregon: AAAI Press, Aug. 2, 1996, pp. 226–231. [Online]. Available: https://cdn.aaai.org/KDD/1996/KDD96-037.pdf.
384. Mautz, D., Plant, C., Böhm, C., "Deep Embedded Cluster Tree," in *2019 IEEE International Conference on Data Mining (ICDM)*, Nov. 2019, pp. 1258–1263. https://doi.org/10.1109/ICDM.2019.00157.

385. McConville, R., Santos-Rodríguez, R., Piechocki, R. J., Craddock, I., "N2D: (Not Too) Deep Clustering via Clustering the Local Manifold of an Autoencoded Embedding," in *2020 25th International Conference on Pattern Recognition (ICPR)*, Jan. 2021, pp. 5145–5152. https://doi.org/10.1109/ICPR48806.2021.9413131.
386. Leiber, C., Bauer, L. G. M., Neumayr, M., Plant, C., Böhm, C., "The DipEncoder: Enforcing Multimodality in Autoencoders," in *Proceedings of the 28th ACM SIGKDD Conference on Knowledge Discovery and Data Mining*, New York, NY, USA: Association for Computing Machinery, Aug. 14, 2022, pp. 846–856. https://doi.org/10.1145/3534678.3539407.
387. Leiber, C., Bauer, L. G. M., Schelling, B., Böhm, C., Plant, C., "Dip-based Deep Embedded Clustering with k-Estimation," in *Proceedings of the 27th ACM SIGKDD Conference on Knowledge Discovery & Data Mining*, New York, NY, USA: Association for Computing Machinery, Aug. 14, 2021, pp. 903–913. https://doi.org/10.1145/3447548.3467316.
388. Ren, Y., Wang, N., Li, M., Xu, Z., "Deep density-based image clustering," *Knowledge-Based Systems*, vol. 197, p. 105 841, Jun. 7, 2020. https://doi.org/10.1016/j.knosys.2020.105841.
389. Moradi Fard, M., Thonet, T., Gaussier, E., "Deep *k*-Means: Jointly clustering with *k*-Means and learning representations," *Pattern Recognition Letters*, vol. 138, pp. 185–192, Oct. 1, 2020. https://doi.org/10.1016/j.patrec.2020.07.028.
390. Xie, J., Girshick, R., Farhadi, A., "Unsupervised deep embedding for clustering analysis," in *Proceedings of the 33rd International Conference on International Conference on Machine Learning—Volume 48*, New York, NY, USA: JMLR.org, Jun. 19, 2016, pp. 478–487.
391. Gove, R., "Automatic Narrative Summarization for Visualizing Cyber Security Logs and Incident Reports," *IEEE Transactions on Visualization and Computer Graphics*, vol. 28, pp. 1182–1190, 2022. https://doi.org/10.1109/TVCG.2021.3114843.
392. Milosevic, N., Dehghantanha, A., Choo, K.-K. R., "Machine learning aided Android malware classification," *Computers & Electrical Engineering*, vol. 61, pp. 266–274, Jul. 1, 2017. https://doi.org/10.1016/j.compeleceng.2017.02.013.
393. Xu, J., Xu, B., Wang, P., Zheng, S., Tian, G., Zhao, J., Xu, B., "Self-Taught convolutional neural networks for short text clustering," *Neural Networks*, vol. 88, pp. 22–31, Apr. 1, 2017. https://doi.org/10.1016/j.neunet.2016.12.008.
394. Anastasiadis, M., Aivatoglou, G., Spanos, G., Voulgaridis, A., Votis, K., "Combining text analysis techniques with unsupervised machine learning methodologies for improved software vulnerability management," in *2022 IEEE International Conference on Cyber Security and Resilience (CSR)*, Jul. 2022, pp. 273–278. https://doi.org/10.1109/CSR54599.2022.9850314.
395. Rendón, E., Abundez, I., Arizmendi, A., Quiroz, E. M., "Internal versus external cluster validation indexes," *International Journal of computers and communications*, vol. 5, pp. 27–34, 2011.
396. Rosenberg, A., Hirschberg, J., "V-Measure: A Conditional Entropy-Based External Cluster Evaluation Measure," in *Proceedings of the 2007 Joint Conference on Empirical Methods in Natural Language Processing and Computational Natural Language Learning (EMNLP-CoNLL)*, Prague, Czech Republic: Association for Computational Linguistics, Jun. 2007, pp. 410–420.

397. Amigó, E., Gonzalo, J., Artiles, J., Verdejo, F., "A comparison of extrinsic clustering evaluation metrics based on formal constraints," *Information Retrieval*, vol. 12, pp. 461–486, Aug. 1, 2009. https://doi.org/10.1007/s10791-008-9066-8.
398. Shu, R., Xia, T., Williams, L., Menzies, T., "Better Security Bug Report Classification via Hyperparameter Optimization." (May 16, 2019).
399. Almeida, T., Hidalgo, J., *SMS Spam Collection*, UCI Machine Learning Repository, Jun. 21, 2012. https://doi.org/10.24432/C5CC84.
400. Hess, S., Duivesteijn, W., "K Is the Magic Number-Inferring the Number of Clusters Through Nonparametric Concentration Inequalities," in *Machine Learning and Knowledge Discovery in Databases*, U. Brefeld, E. Fromont, A. Hotho, A. Knobbe, M. Maathuis, and C. Robardet, Eds., Cham: Springer International Publishing, 2020, pp. 257–273. https://doi.org/10.1007/978-3-030-46150-8_16.
401. Maurus, S., Plant, C., "Skinny-dip: Clustering in a Sea of Noise," in *Proceedings of the 22nd ACM SIGKDD International Conference on Knowledge Discovery and Data Mining*, New York, NY, USA: Association for Computing Machinery, Aug. 13, 2016, pp. 1055–1064.https://doi.org/10.1145/2939672.2939740.
402. Nielsen, F., "Hierarchical Clustering," in *Introduction to HPC with MPI for Data Science*, F. Nielsen, Ed., Cham: Springer International Publishing, 2016, pp. 195–211. https://doi.org/10.1007/978-3-319-21903-5_8.
403. Ankerst, M., Breunig, M. M., Kriegel, H.-P., Sander, J., "OPTICS: Ordering points to identify the clustering structure," *ACM SIGMOD Record*, vol. 28, pp. 49–60, Jun. 1, 1999.https://doi.org/10.1145/304181.304187.
404. Li, Z., Zhang, X., Zhang, Y., Long, D., Xie, P., Zhang, M., "Towards General Text Embeddings with Multi-stage Contrastive Learning." (Aug. 7, 2023).
405. Morris, J. X., Rush, A. M., "Contextual Document Embeddings." (Nov. 8, 2024).
406. Liu, Z., Zhao, C., Fedorov, I., Soran, B., Choudhary, D., Krishnamoorthi, R., Chandra, V., Tian, Y., Blankevoort, T., "SpinQuant: LLM quantization with learned rotations." (Oct. 7, 2024).
407. Jiang, A. Q., Sablayrolles, A., Mensch, A., Bamford, C., Chaplot, D. S., Casas, D., Bressand, F., Lengyel, G., Lample, G., Saulnier, L., Lavaud, L. R., Lachaux, M.-A., Stock, P., Scao, T. L., Lavril, T., Wang, T., Lacroix, T., Sayed, W. E., "Mistral 7B." (Oct. 10, 2023).
408. Zhang, D., Li, J., Zeng, Z., Wang, F., "Jasper and Stella: Distillation of SOTA embedding models." (Jan. 23, 2025).
409. Moreira, G. d. S. P., Osmulski, R., Xu, M., Ak, R., Schifferer, B., Oldridge, E., "NV-Retriever: Improving text embedding models with effective hardnegative mining." (Jul. 22, 2024).
410. Zhang, X., Zhang, Y., Long, D., Xie, W., Dai, Z., Tang, J., Lin, H., Yang, B., Xie, P., Huang, F., Zhang, M., Li, W., Zhang, M., "mGTE: Generalized Long-Context Text Representation and Reranking Models for Multilingual Text Retrieval." (Oct. 14, 2024).
411. Peters, F., Tun, T. T., Yu, Y., Nuseibeh, B., "Text Filtering and Ranking for Security Bug Report Prediction," *IEEE Transactions on Software Engineering*, vol. 45, pp. 615–631, 2019. https://doi.org/10.1109/TSE.2017.2787653.
412. Wu, X., Zheng, W., Xia, X., Lo, D., "Data Quality Matters: A Case Study on Data Label Correctness for Security Bug Report Prediction," *IEEE Transactions on Software Engineering*, vol. 48, pp. 2541–2556, 2022. https://doi.org/10.1109/TSE.2021.3063727.

MIX
Papier aus verantwortungsvollen Quellen
Paper from responsible sources
FSC® C105338

If you have any concerns about our products,
you can contact us on
ProductSafety@springernature.com

In case Publisher is established outside the EU,
the EU authorized representative is:
Springer Nature Customer Service Center GmbH
Europaplatz 3, 69115 Heidelberg, Germany

Printed by Libri Plureos GmbH
in Hamburg, Germany